W9-CBE-116

FIVE

BILLION

YEARS

OF

SOLITUDE

CURRENT

FIVE
BILLION
YEARS
OF
SOLITUDE

THE SEARCH FOR LIFE
AMONG THE STARS

LEE BILLINGS

CURRENT

CURRENT

Published by the Penguin Group
Penguin Group (USA) LLC
375 Hudson Street
New York, New York 10014

USA | Canada | UK | Ireland | Australia | New Zealand | India | South Africa | China
penguin.com
A Penguin Random House Company

First published by Current, a member of Penguin Group (USA) LLC, 2013

Copyright © 2013 by Lee Billings
Penguin supports copyright. Copyright fuels creativity, encourages diverse voices, promotes free speech,
and creates a vibrant culture. Thank you for buying an authorized edition of this book and for
complying with copyright laws by not reproducing, scanning, or distributing any part of it in any form
without permission. You are supporting writers and allowing Penguin to continue to publish
books for every reader.

LIBRARY OF CONGRESS CATALOGING-IN-PUBLICATION DATA
Billings, Lee, author.
Five billion years of solitude : the search for life among the stars / Lee Billings.
pages cm
Includes bibliographical references and index.
ISBN 978-1-61723-006-6
1. Life on other planets.   2. Extrasolar planets.   I. Title.

QB54.B54 22013
576.8'39—dc23   2013017672

Printed in the United States of America
1   3   5   7   9   10   8   6   4   2

Set in Electra LT Std
Designed by Sabrina Bowers

While the author has made every effort to provide accurate telephone numbers, Internet addresses, and
other contact information at the time of publication, neither the publisher nor the author assumes
any responsibility for errors or for changes that occur after publication. Further, publisher does not have
any control over and does not assume any responsibility for author or third-party Web sites or their content.

To Mike and Pam, Bruce and Jo, Melissa,
and all those with the courage to keep looking up

CONTENTS

FIVE
BILLION
YEARS
OF
SOLITUDE

INTRODUCTION

Here on Earth we live on a planet that is in orbit around the Sun.
The Sun itself is a star that is on fire and will someday burn up,
leaving our solar system uninhabitable. Therefore we must build a
bridge to the stars, because as far as we know, we are the only
sentient creatures in the entire universe. . . . We must not fail in
this obligation we have to keep alive the only meaningful life we
know of.

—WERNHER VON BRAUN,
ARCHITECT OF NASA'S APOLLO PROGRAM,
AS RECALLED BY TOM WOLFE

This story properly begins 4.6 billion years ago, with the birth of our solar system from a cloud of cold hydrogen and dust several light-years wide. The cloud was but a wisp from a much larger mass of primordial gas, a stellar nursery manufacturing massive stars destined to explode as supernovae. One by one, the giant stars popped off like firecrackers, ejecting heavy elements that sizzled with radioactivity as they rode shock waves through the murk like so much scattered confetti. One of those enriching shock waves may have compressed the cloud, our cloud, in its passage. The cloud became dense enough for gravity to seize control, and it collapsed in on itself. Most of its material fell to its

—
1

center to form a hot, simmering protostar. Eventually, the protostar gained enough mass to kindle a thermonuclear fire at its core, and the Sun began to shine. What was left of the cloud settled around the new-born star in a turbulent, spinning disk of incandescent vapor.

Microscopic grains of metal, rock, ice, and tar rained out from the whirling disk as it slowly cooled. The grains swirled through the disk for millennia, occasionally colliding, sometimes sticking together, gradually glomming into ever-larger objects. First came millimeter-scale beads, then centimeter-scale pebbles, then meter-scale boulders, and finally kilometer-scale orbiting mountains called "planetesimals." The planetes-imals continued to collide, forming larger masses of ice, rock, and metal that grew with each impact. Within a million years, the planetesimals had grown into hundreds of Moon-size embryos, protoplanets that through violent collisions grew larger still, until they became full-fledged worlds.

After perhaps one hundred million years of further collisions, the embryos in the inner solar system had combined to make Earth and the other rocky planets. The inner worlds were likely bone-dry, their water and other volatiles blowtorched away by the intense light of the newborn Sun. In the outer solar system, freezing temperatures locked the volatiles in ice. The ices provided more-solid construction mate-rial, allowing the cores of Jupiter and the other outer planets to rapidly form and sweep up lingering gas within the disk in only a few million years. As they grew, the giant planets created zones of instability where embryos could not assemble, leaving behind pockets of primordial planetesimals and bands of shattered rock and metal. These remnants are the asteroids. The giant planets also catapulted many icy planetesi-mals far out into the solar hinterlands, to drift in the dark out beyond the orbit of present-day Pluto. When jostled by perturbing planets, ga-lactic tides, or close-passing stars, those icy outcasts fall back toward the Sun as comets.

Finally, sometime between 3.8 and 4 billion years ago, a complex, chaotic, hazily understood series of gravitational interactions between the giant planets stirred up most of the outer solar system, sending barrages of asteroids and comets hurtling sunward to pound the dry, rocky inner worlds. This event is called the "Late Heavy Bombardment," and was the last gasp of planet formation. We observe its effects in the cratered surface of the Moon, and also in the rain that has eroded its geographic scars from our own planet—much of Earth's water seems to have arrived during the Bombardment, express-delivered from the outer solar system. Afterward, Earth's crust had partially melted, and its original atmosphere had been mostly swept away. But as those first torrential rains fell from the steam-filled sky, our planet gained the gift of oceans. Slowly, the Earth cooled, and gas-belching volcanoes gradually replenished the atmosphere. Soon, perhaps uniquely of all the new-formed worlds of the solar system, ours would somehow come alive.

Slightly less than four billion years later, I was four years old, standing with my mother, father, and sister in our backyard in Jasper, Alabama. It was January 1986, shortly after sunset. My father had built a small bonfire, and we clustered around it against the evening chill, roasting marshmallows as the stars came out overhead. Lower in the sky, just above the treetops, a soft white smear was barely visible. It was Halley's comet, passing near Earth on its trip around the Sun. I remember asking whether I could visit it. I had recently seen the 1974 film adaptation of Saint-Exupéry's *The Little Prince*, and, like the small boy living on an asteroid in the story, I, too, wanted to fly through space to see all the solar system's strange places. "Maybe someday," the answer came. Weeks later, I and the rest of a generation of children would learn that space travel is no fairy tale, watching as NASA's space shuttle *Challenger* broke apart on its way to orbit.

I didn't know then that Halley's comet would not be coming back until far-off 2061, and I was much too young to feel the weight of that date. The comet didn't feel it, either—when it returned, it would be practically unchanged. I, on the other hand, would be nearing my eightieth year on Earth, if I was so lucky. With a great deal more luck, my parents would see it through centenarian eyes.

—

3

When I was ten, after we had moved to Greenville, South Carolina, my mother spent much of one summer teaching illiterate adults to read at a local library. She always brought me along, letting me wander the shelves unsupervised. I began reading enormous amounts of science fiction about alien civilizations and interstellar travel, as well as books about astronomy, which tended to gloss over the possibility of planets and lives beyond our solar system in favor of bigger, flashier things—exploding suns, colliding galaxies, voracious black holes, and the Big Bang. Such was the spirit of the times: for most of the twentieth century, astronomers had been all-consumed by a quest to gaze ever deeper out into space and time, pursuing the fundamental origins and future of existence itself. That quest had revealed one revolutionary insight after another, showing that we lived in but one of innumerable galaxies, each populated by hundreds of billions of stars, all in an expanding universe that began nearly fourteen billion years ago and that might endure eternally. I thrilled at the cosmological creation story but couldn't help but think that it was missing something. Namely, us. Lost somewhere in between the universe's dawn and destiny, a ball of metal, rock, and water called Earth had given birth not only to life, but to sentient beings, creatures with the intellectual capacity to discover their genesis and the technological capability to design their fate. Creatures that, before their sun went dim, might somehow touch the stars. Maybe what had happened once would happen many times, in many places. My father saw the galaxies and stars on the covers of my checked-out library books and bought me a department-store telescope.

Looking through my telescope, I was soon disappointed to learn I couldn't see many of the cosmic fireworks described in the astronomy books, or any evidence for the galactic empires of science fiction. Everything out there looked awfully, deathly quiet. It seemed in all that cosmic space, and thus in the great minds of many learned astronomers, there was paradoxically no room for living beings and their tiny home worlds. Such things were too small to be searched for, too insignificant to be of notice. I kept looking every now and then anyway, half-hoping I might catch a UFO in my viewfinder as it streaked across the sky, or see the bright flashes of some interstellar battle in the twinkling of a star. One day I asked my father whether any planets at all existed around other stars. He thought a moment, and replied that other stars probably had planets, but that no one really knew; none had ever been found, because they were all so far away. After that, most times when I gazed up at the night sky, I would wonder what those planets might look like. Would they be like Earth? Would they have oceans and mountains, coral reefs and grasslands? Would they have cities and farms, computers and radios, telescopes and starships? Would creatures there live and die as we did, or look up and wonder about life's purpose? Would they be lonely? Staring at the trembling stars, I dreamed of worlds I thought I would never see.

By the mid-2000s, I had followed my curiosity into a career in science journalism, where instead of pestering personal friends and acquaintances with my questions, I could simply pester the experts themselves. Answers to some of my earlier questions had emerged over the intervening years: planets proved quite common around other stars, and since the mid-1990s astronomers had found hundreds of them. These worlds were called "exoplanets," and most were far too large and far too near their suns to be hospitable to life as we know it. Using large telescopes on the ground and in space, astronomers had even managed to take pictures of a few that were very hot, very big, and relatively nearby. But other questions remained unaddressed: Were there other Earth-size, Earth-like exoplanets in our galaxy and in the wider universe? Was our situation here on Earth average, or was it instead quite special, even unique? Were we cosmically alone? I decided to write this book when I learned just how soon we might gain answers to some of these seemingly timeless questions.

It was 2007, and I was interviewing the University of California, Santa Cruz, astrophysicist Greg Laughlin for a story. During our chat, Laughlin explained that since exoplanet searches were becoming progressively more sophisticated and capable, there would soon be thousands rather than hundreds of known exoplanets to compare with our own. Astronomy's next big thing, he suggested, would focus not on the edges of space and the beginning of time, but on the nearest stars and the uncharted, potentially habitable worlds they likely harbored. Near the end of our conversation, he guessed that the first Earth-size exoplanets would probably be found within the next five years. He had graphed the year-to-year records for lowest-mass exoplanets, drawing a trend line through the data that suggested an Earth-mass planet would be discovered in mid-2011. It suddenly seemed I had stumbled upon some magnificent secret, hidden in plain view. The more exoplanet-related press releases and papers I read, the more convinced I became that somewhere on Earth there were scientists who would be remembered in history for discovering the first habitable worlds beyond the solar system, and perhaps even the first evidence of extraterrestrial life. Yet they were largely anonymous, utterly unknown to the average person. I wanted to learn more about them, and tell their stories. One by one, I sought them out.

Most welcomed me with open arms, and the ones who didn't still politely tolerated me. Many planned for a bright near-future, one in which they would use great, government-built techno-cathedrals of glass and steel on remote mountaintops and in deep space to wring secrets from the heavens and investigate any promising exoplanets for signs of life. Looking further out in time, some even envisioned our culture eventually escaping Earth entirely to expand into the wider solar system and beyond, driven by a curiosity so insatiable and restless that it would forever propel us outward into the endless immensities of new, far-flung physical frontiers. And yet, as I researched the book, I saw many of their boldest hopes dashed as crucial telescopes and missions were delayed or canceled, deferring all those dreams for generations, if not forever. On the verge of epochal revelations, their work had faltered, but not because of any newfound limitations of celestial physics. Instead, rapid progress in the search for life beyond Earth had succumbed to purely human, mundane failings—negligent organizational stewardship, unsteady and insufficient funding, and petty territorial bickering. Time and time again I felt I was witnessing the planet hunters reach for the stars just as the sky began to fall. And so I became committed to telling not only their personal stories, but also the story of their field, where it came from and where, with a reversal of fortune, it might still go.

The result is the book you now hold in your hands. By necessity, it glosses over or fails to mention numerous discoveries and discoverers that deserve entire shelves of dedicated literature. I hope the knowledgeable reader will forgive my omissions in light of all that this work does encompass. It is a portrait of our planet, revealing how the Earth came to life and how, someday, it will die. It is also a chronicle of an unfolding scientific revolution, zooming in on the ardent search for other Earths around other stars. Most of all, however, it is a meditation on humanity's uncertain legacy.

This book's title, *Five Billion Years of Solitude*, refers to the longevity of life on Earth. Life on this planet has an expiration date, if for no other reason than that someday the Sun will cease to shine. Life emerged here shortly after the planet itself formed some four and a half billion years ago, and current estimates suggest our world has a good half billion years left until its present biosphere of diverse, complex multicellular life begins an irreversible slide back to microbial simplicity. In all this time, Earth has produced no other beings quite like us, nothing else that so firmly holds the fate of the planet in its hands and possesses the power to shape nature to its whim. We have learned to break free of Earth's gravitational chains, just as our ancient ancestors learned to leave the sea. We've built machines to journey to the Moon, travel the breadth of the solar system, or gaze to the edge of creation. We've built others that can gradually cook the planet with greenhouse gases, or rapidly scorch it with thermonuclear fire, bringing a premature end to the world as we know it. There is no guarantee we will use our powers to save ourselves or our slowly dying world and little hope that, if we fail, the Earth could rekindle some new technological civilization in our wake of devastation.

—

In the long view, then, we are faced with a choice, a choice of life or death, a choice that transcends science to touch realms of the spiritual. As precious as the Earth is, we can either embrace its solitude and the oblivion that waits at world's end, or pursue salvation beyond this planetary cradle, somewhere far away above the sky. In our lives we all in some way contribute to this greater choice, either drawing our collective future down to Earth or thrusting it out closer to the stars. Some of the people in this book have devoted themselves to seeking signs of other, wildly advanced galactic cultures, hoping to glimpse our own possible futures via interstellar messages carried on wisps of radio waves or laser light. Others closely study the evolution of Earth's climate over geological time, trying to pin down the limitations of habitability on our own and other worlds. A few have become makers of maps and crafters of instruments, and strive to find the most promising worlds that untold years from now could welcome our distant descendants. All seem to believe that in the fullness of planetary time any human future can only be found far beyond the Earth. You will find their tales, and others, recorded in these pages.

I won't pretend to know what our collective choice will be, how exactly we would embark on such an audacious adventure, or what we would ultimately find out there. I am content to merely have faith that we do, in fact, have a choice. Similarly, I can't suggest that we simply ignore all of our planet's pressing problems by dreaming of escape to the stars. We must protect and cherish the Earth, and each other, for we may never find any other worlds or beings as welcoming. Even if we did, we as yet have no viable way of traveling to them. Here, now, on this lonely planet, is where all our possible futures must begin, and where I pray they will not end.

LEE BILLINGS
NEW YORK CITY, 2013

CHAPTER 1

Looking for Longevity

On a hillside near Santa Cruz, California, a split-level ranch house sat in a stand of coast redwoods, the same color as the trees. Three small climate-controlled greenhouses nestled alongside the house next to a diminutive citrus grove, and a satellite dish was turned to the heavens from the manicured back lawn. Sunlight filtered into the living room through a cobalt stained-glass window, splashing oceanic shades across an old man perched on a plush couch. Frank Drake looked blue. He leaned back, adjusted his large bifocal glasses, folded his hands over his belly, and assessed the fallen fortunes of his chosen scientific field: SETI, the search for extraterrestrial intelligence.

"Things have slowed down, and we're in bad shape in several

ways," Drake rumbled. "The money simply isn't there these days. And we're all getting old. A lot of young people come up and say they want to be a part of this, but then they discover there are no jobs. No company is hiring anyone to search for messages from aliens. Most people don't seem to think there's much benefit to it. The lack of interest is, I think, because most people don't realize what even a simple detection would really mean. How much would it be worth to find out we're not alone?" He shook his head, incredulous, and sunk deeper in the couch.

Besides a few extra wrinkles and pounds, at eighty-one years old Drake was scarcely distinguishable from the young man who more than half a century earlier conducted the first modern SETI search. In 1959, Drake was an astronomer at the National Radio Astronomy Observatory (NRAO) in Green Bank, West Virginia. He was only twenty-nine then, lean and hungry, yet he already possessed the calm self-assurance and silver hair of an elder statesman. At work one day, Drake began to wonder just what the site's newly built 85-foot-wide radio dish was capable of. He performed some back-of-the-envelope calculations based on the dish's sensitivity and transmitting power, then probably double-checked them with a growing sense of glee. Drake's figuring showed that if a twin of the 85-footer existed on a planet orbiting a star only a dozen light-years away, it could transmit a signal that the dish in Green Bank could readily receive. All that was needed to shatter Earth's cosmic loneliness was for the receiving radio telescope to be pointed at the right part of the sky, at the right time, listening at the right radio frequency.

"That was true then, and it's true today," Drake told me. "Right now there could well be messages from the stars flying right through this room. Through you and me. And if we had the right receiver set up properly, we could detect them. I still get chills thinking about it."

It didn't take long for Drake to discuss the wild prospect with his superiors at NRAO. They granted him a small budget to conduct a simple search. During the spring of 1960, Drake periodically pointed the 85-footer at two nearby Sun-like stars, Tau Ceti and Epsilon Eridani, to listen for alien civilizations that might be transmitting radio signals toward Earth. Drake called the effort "Project Ozma," after the princess who ruled over the imaginary Land of Oz in Frank Baum's popular series of children's books. "Like Baum, I, too, was dreaming of a land far away, peopled by strange and exotic beings," he would later write.

Project Ozma recorded little more than interstellar static, but still inspired a generation of scientists and engineers to begin seriously considering how to discover and communicate with technological civilizations that might exist around other stars. Over the years, astronomers used radio telescopes around the world to conduct hundreds of searches, looking at thousands of stars on millions of narrowband radio frequencies. But not one delivered unassailable evidence of life, intelligence, or technology beyond our planet. The silence of the universe was unbroken. And so for more than fifty years Drake and his disciples played the roles of not only scientists but also salespeople. For the entirety of the discipline's existence, SETI groups had been searching nearly as ardently for sources of funding as they had for signals from extraterrestrials.

Early on, governments were quite interested—SETI was briefly one of the scientific arenas in which the United States and the Soviet Union grappled during the Cold War. What better propaganda victory could there be than to act as humanity's ambassador to another cosmic civilization? What invaluable knowledge might be gained—and exploited—from communication between the stars? In 1971, a prestigious NASA commission concluded that a full-bore search for alien radio transmissions from stars within a thousand light-years of Earth would require an array of giant radio telescopes with a total collecting area of between 3 and 10 square kilometers, built at a cost of about $10 billion. Politicians and taxpayers balked at the price tag, and SETI began its long descent from political favor. The trend of null results stretched out over decades, and already scarce and fickle federal funding for American SETI efforts progressively dwindled. A glimmer of hope emerged in 1992, when NASA launched an ambitious new SETI program, but congressional backlash shuttered the project the following year. Since 1993, not a single federal dollar had directly sponsored the search for radio messages from the stars. Drake and a group of his disciples had suspected what was coming, so in 1984 they formed a nonprofit research organization, the SETI Institute, to more easily solicit financial support from both the public and private sectors. Headquartered in Mountain View, California, the SETI Institute began to thrive in the mid-1990s through a combination of research grants and private donations from starry-eyed and newly wealthy Silicon Valley technologists. Drake served as the Institute's president from its founding until 2000, before transitioning into an active retirement a couple of years after the turn of the millennium.

By 2003, the Institute had secured $25 million in funding from Paul Allen, the billionaire cofounder of Microsoft, to build an innovative new instrument, the Allen Telescope Array (ATA), in a bowl-shaped desert valley some 185 miles north of San Francisco. Rather than construct a smaller number of gigantic (and gigantically expensive) dishes, the Institute would save money by building larger numbers of smaller dishes. Drake had spearheaded much of the ATA's design. Three hundred fifty 6-meter dishes would act together as one extremely sensitive radio telescope, monitoring an area of sky nearly five times larger than the full Moon on a wide range of frequencies. Allen's millions, along with $25 million more from other sources, were sufficient to build the ATA's first forty-two dishes, which were completed in 2007. Significant funds to operate the fledgling ATA came from California state funding and federal research grants to the Radio Astronomy Laboratory at the University of California, Berkeley, which jointly ran the ATA with the Institute. Though only partially completed, the ATA still functioned well enough to support a SETI effort as well as a significant amount of unrelated radio astronomy research. It operated on an annual budget of approximately $2.5 million—at least until 2011, when funding shortfalls forced the entire facility into hibernation.

As I spoke with Drake in his home in June 2011, weeds were already growing up around the idle dishes at the shuttered ATA. Only a skeleton crew of four Institute employees remained attached to the facility, merely to ensure it wouldn't fall into irreparable disarray. The ATA would not restart operations until December, buoyed by a brief flurry of small donations. The money raised was sufficient to fund only another few months of operations. The Institute was seeking a partnership with the U.S. Air Force, which later purchased time on the ATA to monitor "space junk"—cast-off rocket stages, metal bolts, and other debris that can strike and damage spacecraft. But that funding, too, proved only temporary, and time spent surveying space junk was time sucked away from the ATA's SETI-centric goals. Unless more wealthy patrons swooped in with heavyweight donations, the ATA had very little hope of reaching its original target of 350 dishes—and during the long recession after the 2008 turmoil in the global financial system, potential donors were proving at least as elusive as any broadcasting aliens. Drake's greatest dream seemed to be collapsing.

Aside from political and economic difficulties, there was another factor in SETI's decline that was at once more scientific and particularly ironic: the rise of exoplanetology, a field devoted to the discovery and study of exoplanets, planets orbiting stars other than our Sun. Beginning in the early 1990s, as radio telescopes intermittently swept the skies for messages from extraterrestrials, a revolution occurred in astronomy. Observers using state-of-the-art equipment began finding exoplanets with clockwork regularity. The first worlds discovered were "hot Jupiters," bloated and massive gas-giant worlds orbiting inhospitably close to their stars. But as planet-hunting techniques grew more sophisticated, the pace of discovery quickened, and ever-smaller, more life-friendly worlds began to turn up. Twelve exoplanets were discovered in 2001, all of which were hot Jupiters. Twenty-eight were found in 2004, including several as small as Neptune. The year 2010 saw the discovery of more than a hundred worlds, a handful of which were scarcely larger than Earth. By early 2013, a single NASA mission, the

Kepler Space Telescope, had discovered more than 2,700 likely exoplanets. A small fraction of Kepler's finds were the same size as or smaller than Earth and orbited in regions around stars where life as we know it could conceivably exist. Emboldened, astronomers earnestly discussed building huge space telescopes to seek signs of life on any habitable worlds around nearby stars.

When the ATA briefly came back online in December of 2011, it began to survey those promising Kepler candidates for the radio chatter of any talkative aliens who might live there. No signals were detected before the ATA was sent back into hibernation, starved once again for money. SETI's half century of null results could not be further from the ongoing exoplanet boom, where sensational discoveries could lead to media fame, academic stardom, and plentiful funding for researchers and institutions. For those interested in extraterrestrial life, exoplanetology, not SETI, was the place to be. As the search for Earth-like planets came to a boil, SETI was being frozen out of the scientific world.

When I asked Drake if we were witnessing the end of SETI, his blue eyes twinkled behind a knowing Cheshire Cat grin. "Oh no, not at all. This, I think, has been just the beginning. People presume we've been somehow monitoring the entire sky at all frequencies, all the time, but we haven't yet been able to do any of those things. The fact is, all the SETI efforts to date have only closely examined a couple thousand nearby stars, and we're only just now learning which of those might have promising planets. . . . Even if we have been pointed in the right direction and listening at the right frequency, the probability of a message being beamed at us while we're looking is certainly not very large. We've been playing the lottery using only a few tickets."

Drake's confidence that there are other life forms out there at all had its roots in a private meeting that took place shortly after Project Ozma.

In 1961, J. P. T. Pearman of the U.S. National Academy of Sciences approached Drake to help convene a small, informal SETI conference at NRAO's Green Bank observatory. The core purpose of the meeting, Pearman explained, was to quantify whether SETI had any reasonable chance of successfully detecting civilizations around other stars. The "Green Bank conference" was held November 1–3, 1961.

The invite list was star-studded and short. Besides Drake and Pearman, three Nobel laureates attended. The chemist Harold Urey and the biologist Joshua Lederberg had both won Nobel Prizes in their fields, Urey for his discovery of deuterium, a heavier isotope of hydrogen, and Lederberg for his discovery that bacteria could mate and swap genetic material. Both were early practitioners in the still-nascent field of astrobiology, the study of life's origins and manifestations in space. Urey was particularly interested in the prebiotic chemistry of the ancient Earth, and Lederberg worked to define how alien life on a distant planet might be remotely detected. As the conference was underway, one of the guests, the chemist Melvin Calvin, was awarded a Nobel for his elucidation of the chemical pathways underlying photosynthesis.

The other attendees were only slightly less celebrated. The physicist Philip Morrison had coauthored a 1959 paper advocating a SETI program just like the one Drake undertook in 1960. Dana Atchley was an expert in radio communications systems and president of Microwave Associates, Inc., a company that had donated equipment for Drake's search. Bernard Oliver was vice president of research at Hewlett-Packard, and already an avid SETI supporter, having earlier traveled to Green Bank to witness Drake's first search. The Russian-born American astronomer Otto Struve, the director of Green Bank observatory, invited one of his star pupils, a soft-spoken NASA researcher named Su-Shu Huang. Struve was a legendary optical astronomer, and one of the first who seriously considered how to find planets orbiting other stars. He and Huang had worked together studying how a star's mass and luminosity could affect the habitability of any orbiting planets. The neuroscientist John Lilly came to Green Bank to present his ideas on interspecies communication, based on his controversial experiments with captive bottlenose dolphins. A dark-haired and brilliant twenty-seven-year-old astronomy postdoc named Carl Sagan was, at the time, the youngest and arguably least distinguished name on the guest list. Lederberg, one of Sagan's mentors, had invited him.

It fell to Drake to arrange the agenda. A few days before the conference began, he sat down at his desk with pencil and paper and tried to categorize all the key pieces of information needed to estimate the number, N, of detectable advanced civilizations that might currently exist in our galaxy. He began with the fundamentals: surely a civilization could only emerge on a habitable planet orbiting a stable, long-lived star. Drake reasoned that the average rate of star formation in the Milky Way, R, thus placed a rough upper limit on the creation of new cradles for cosmic civilizations. Some fraction of those stars, f_p, would actually form planets, and some number of those planets, n_e, would be suitable for life. From astrophysics and planetary science, Drake's musing entered into the field of evolutionary biology: some fraction of those habitable planets, f_l, would actually blossom into living worlds, and some fraction of those living worlds, f_i, would give birth to intelligent, conscious beings. As his considerations shifted to the rarefied realms of social science, Drake became restless. He sensed he was nearing the end of his categories and the outer limits of reasonable speculation. He doggedly forged ahead. The fraction of intelligent extraterrestrials who developed technologies that could communicate their existence across interstellar distances was f_c, and the average longevity of a technological society was L.

Longevity was important, Drake believed, because of the Milky Way's sheer size and immense age, and the inconvenient fact that nothing seemed able to travel through space faster than the speed of light. Approximately 100,000 light-years wide, and thought to be almost as old as the universe itself, our galaxy presented a huge volume of space and time in which other cosmic civilizations could pop up. If, for example, the average lifetime of an advanced technological society was a few hundred years, and two such societies emerged simultaneously around stars a thousand light-years apart, they would have essentially no chance of making contact before various forces brought the communicative phases of their empires to an end. Even if one somehow discovered the other, and beamed a message toward that distant star, by the time the message arrived a millennium later, the society that sent the message would no longer exist.

If one were to multiply all of Drake's factors together using plausible figures, conceivably a ballpark estimate of N would emerge. The terms were interdependent; if any one of them had a vanishingly low value, the resulting N, the estimated number of detectable technological civilizations at large in the Milky Way, would drop precipitously. Strung together, they formed an equation of sorts that, if it did not yield an accurate estimate of contemporaneous cosmic civilizations, at least helped quantify humanity's cosmic ignorance.

On the morning of November 1, after the guests were seated and sipping coffee in a small lounge in the NRAO residence hall, Drake rose and strode forward to present what he'd come up with. But rather than address the room from the central lectern, he kept his back turned and scratched out his lengthy figure on a nearby blackboard. When he put down the chalk and stepped aside, the board read:

$$N = R\, f_p\, n_e\, f_l\, f_i\, f_c\, L$$

That string of letters has come to be known as the "Drake equation." Though Drake had intended it only to guide the next three days of the Green Bank meeting, the equation and its plausible values would come to dominate all subsequent SETI discussions and searches.

At the time, only one term, R, the rate of star formation, was reasonably constrained. Astronomers had already closely studied several

star-forming regions in the Milky Way. Based on that data, the astronomers in the group quickly pegged R at a conservative value of at least
one per year within our galaxy. They also chose to focus on Sun-like
stars. Stars much larger than our own were also far more luminous,
and burned out in only tens or hundreds of millions of years, leaving
little time for complex life to arise on any orbiting planets. Stars much
smaller than the Sun were far more parsimonious with their nuclear
fuel, and could weakly shine for hundreds of billions of years. But to be
sufficiently warmed by that dim light a planet would need to be perilously close to the star, where stellar flares and gravitational tides could
wreak havoc on a biosphere. Sun-like stars struck a balance between
the two extremes, steadily shining for several billions of years with sufficient luminosity for habitable planets to exist far removed from stellar
fireworks.

In 1961, no planets beyond our solar system were yet known, so the
estimate of f_p relied only on indirect evidence. It emerged from a discussion between Struve and Morrison. Struve had performed pioneering
work decades earlier, measuring the rotation rates of different types of
stars. He found that the very hot, very massive stars larger than our Sun
spun very fast, while stars like our own, as well as those that were smaller
and cooler, spun more slowly. The difference, Struve thought, was that
spinning planets accompanied the stars more like our Sun, sapping the
stars' angular momentum and reducing their spin rates. However,
roughly half of the known Sun-like stars were in binary systems, co-
orbiting with a companion star that could also affect their spin. In such
a system the pull of each star upon the other, it was thought, might disrupt the process of planet formation. Struve speculated that only the
other half, the singleton suns, would be likely to form planets. He was
so convinced that planets were common around Sun-like stars that
almost a decade earlier, in 1952, he had published a paper laying out
two observational strategies to find them, presaging the exoplanet boom
by a half century. Struve's estimate that half of all Sun-like stars had
planets was too high for Morrison, who guessed that even around many

solitary stars only scattered asteroids and comets would form. He thought f_p might be as low as one-fifth.

Next, the group turned to n_e, the number of habitable planets per system. Huang and Struve marshaled their years of work together to posit that our own solar system's architecture was typical, with a large number of planets in a wide distribution of orbits. In any system, they suggested, at least one world would fall within Huang's "habitable zone," the broadly defined circumstellar region where liquid water could exist on a planet's surface. Sagan concurred, and pointed out that abundant greenhouse gases in a planet's atmosphere could act to warm an otherwise frigid planet, greatly extending the habitable zone's expanse. Looking to our own solar system, the group focused on scorched Venus and frigid Mars, two borderline worlds that, if they possessed moderately different atmospheric compositions, would likely be quite Earth-like indeed. Accounting for Sagan's proposed greenhouse extension of Huang's habitable zone, the attendees decided that a planetary system would likely harbor anywhere from one to five planets suitable for life. They pegged n_e at somewhere between one and five. Of course, billions of habitable planets could exist in the galaxy and none other than Earth might be inhabited, if life's origin was a cosmic fluke.

As the discussion turned to the value of f_l, the number of habitable planets that gave birth to life, it entered Urey and Calvin's realm of expertise. In 1952, Urey had teamed with one of his graduate students, Stanley Miller, to investigate the origins of life on the primordial Earth, where geothermal heating, lightning strikes, and ultraviolet light from the tempestuous young Sun would have suffused the environment with useful energy. The duo decided to run a modest electric current through a sealed vessel of hydrogen, methane, ammonia, and water vapor—a mixture of gases thought at the time to mimic Earth's ancient atmosphere. After only a week the Urey-Miller experiment had synthesized a "primordial soup" of organic compounds—sugars, lipids, and even amino acids, which are the building blocks of proteins. Acting over millions of years on a planetary scale, such reactions could easily

---

synthesize the organic ingredients for life from inorganic chemical precursors. On our own planet, the fossil record suggested that life must have already been thriving only a few hundred million years after our planet cooled from its formation; it seemed to have appeared as soon as it possibly could.

Calvin argued forcefully that on geological timescales the emergence of simple, single-celled life was a certainty on any habitable world. Sagan noted that astronomers had already detected hydrogen, methane, ammonia, and water in clouds of interstellar gas and dust, and that even some varieties of meteorites were proving to be rich in organic compounds. All this suggested that planets with atmospheres similar to the early Earth's would be common outcomes of planet formation, Sagan said. And, since the laws of physics and chemistry were everywhere the same, when warmed by their stars' light these worlds would become enriched with life's organic building blocks. Through innumerable iterations and permutations of organic compounds in the primordial soup, crude catalytic enzymes and self-replicating molecules would gradually emerge, and life's genesis would be at hand. The rest of the group agreed: given hundreds of millions or billions of years, single-celled life would likely spring up on each and every habitable world, yielding an f_l value of one.

When the time came to discuss f_i, the fraction of habitable, life-bearing planets that develop intelligent inhabitants, Lilly discussed his experiments with captive dolphins on the island of Saint Thomas in the Caribbean. Lilly began by noting that the brain of a dolphin was larger than that of a man, with similar neuron density and a richer variety of cortical structure. He recounted his various attempts to communicate with the dolphins in their own language of clicks and whistles, and told stories of dolphins rescuing sailors lost at sea. He focused on one case in which two of his captive dolphins had acted together to rescue a third from drowning when it became fatigued in the cold water of a swimming pool. The chilled dolphin had let out two sharp whistles in an apparent call for help, spurring the two rescuers to chatter together, form a rescue plan, and save their distressed companion. The display convinced Lilly that dolphins were a second terrestrial intelligence contemporaneous with humans, capable of complex communication, future planning, empathy, and self-reflection.

Morrison broadened the discussion by introducing the concept of convergent evolution, the tendency for natural selection to sculpt creatures from very different evolutionary lineages into common forms to fit shared environments and ecological niches. Hence, fish such as tuna or sharks shared a streamlined body form with mammalian dolphins, and features such as eyes and wings had independently evolved across the animal kingdom several times. Perhaps, Morrison said, intelligence was another example of convergent evolution, and had emerged not only in humans and dolphins but also in other primates and cetaceans, such as whales and now-extinct Neanderthals. Like eyes or wings, intelligence might be an extremely successful adaptation that would emerge time and time again in a planetary environment—provided life first made the fundamental evolutionary leap from simple solitary cells to complex multicellular organisms. Moved by Morrison's arguments, the Green Bank scientists optimistically placed the value of f_i at one.

Morrison also proved decisive in framing the Green Bank debate over the two final and most nebulous terms of Drake's equation: f_c, the fraction of intelligent creatures who would develop societies and technologies capable of interstellar communication, and L, the average longevity of an advanced technological civilization. He first noted that while creatures like dolphins and whales might well be intelligent, in their current aquatic forms they seemed destined for cosmic invisibility: supposing they had language and culture, they still lacked a way of assembling or using even relatively simple tools and machines. None of the attendees could easily imagine any cetacean civilization ever building anything like a radio telescope or a television broadcast antenna. But on land, Morrison said, history suggested that the emergence of technological societies might be another convergent phenomenon. The early civilizations of China, the Middle East, and the Americas all arose independently and generally followed similar lines of development.

And yet, the drivers of social change and technological progress were not at all clear. Despite China's development of technologies such as gunpowder, compasses, paper, and the printing press hundreds of years before Europeans did, China experienced nothing equivalent to the European Renaissance and the successive scientific and industrial revolutions. When Spanish and Portuguese explorers, rather than the Chinese, used great ocean-faring ships to discover the Americas, they found indigenous civilizations using Stone Age technology that was no match for European steel and gunpowder. Sending ships across oceans or messages between the stars appeared to be a matter not only of technological prowess, but also of choice. Whether any given technological culture would attempt interstellar communication seemed unpredictable. Facing a somewhat arbitrary decision, the Green Bank attendees eventually guessed that between one-fifth and one-tenth of intelligent species would develop the capabilities and intentions to search for and signal other cosmic civilizations. That left only L, the typical lifetime of technological civilizations, for the group to consider.

During a break in the proceedings, Drake noticed something that made him suspect his equation could be substantially streamlined: Three of the equation's seven terms (R, f_l, f_i) appeared to be equal to one, and hence would have little effect on the product N, the number of detectable civilizations in our galaxy. Similarly, the plausible values of the other three terms (f_p, n_e, f_c) could easily cancel each other out. For instance, the group had guessed that the average number of habitable planets per system, n_e, was between one and five, and that f_p, the fraction of stars with planets, was between one-half and one-fifth. If the value of n_e was actually two, and f_p's value was one-half, multiplied together the result was one, and N was scarcely affected. After considering the best evidence that was available, some of the brightest scientific minds on planet Earth had concluded that the universe, on balance, was a rather hospitable place, one that surely must be overflowing with living worlds. It stood to reason that, on other planets circling other suns, other curious minds gazed at their night skies wondering if they, too, were alone. And yet, Drake announced, more than the number of stars, or the number of habitable planets, or how often life, intelligence, and high technology emerged, what he suspected really controlled the number of technological civilizations currently extant in the cosmos was almost solely their longevity. $N=L$.

The thought made Morrison shudder. Of all the Green Bank attendees, he alone could viscerally appreciate just how fleeting our modern era might be. He had worked on the Manhattan Project during World War II, and had witnessed the detonation of the first atomic bomb, at Alamogordo, New Mexico, on July 16, 1945. A month later, on the South Pacific island of Tinian, Morrison had personally assembled and armed an atomic bomb that was later dropped on the Japanese city of Nagasaki. Tens of thousands of civilians were incinerated in the bomb's fireball, and tens of thousands more died slowly from secondary burns and exposure to radioactive fallout, all from the nuclear fission of about two pounds of plutonium. When Japan's surrender drew the war to a close, Morrison was among a contingent of American scientists who toured the cities of Hiroshima and Nagasaki to evaluate up close the devastation wrought by atomic warfare. Shortly after, he became a vocal proponent of nuclear disarmament, but it was too late. The Soviet Union had already begun a crash program to develop atomic bombs, and would successfully test its first nuclear weapon in 1949. In the ensuing arms race both the United States and the Soviet Union succeeded in harnessing the far more powerful process of thermonuclear fusion, squeezing the destructive force of hundreds of Nagasakis into individual bombs. The resulting arsenals of thermonuclear weapons were more than adequate to extinguish hundreds of millions of lives in a single nuclear exchange. Those who survived such a nuclear holocaust would face a severely damaged planetary biosphere and a world plunged into a new Dark Age. Less than a year after the Green Bank proceedings, the Cuban missile crisis would bring the world to the brink of thermonuclear war, and as time marched on, more and more nations successfully weaponized the power of the atom. Humans had developed a global society, radio telescopes, and interplanetary rockets at roughly the same time as weapons of mass destruction.

If it could happen here, Morrison gloomily suggested, it could happen anywhere. Perhaps all societies would proceed on similar trajectories, becoming visible to the wider cosmos at roughly the same moment they gained an ability to destroy themselves. In fact, he went on, running the numbers in his steel-trap mind, if the average civilization endured only a decade before passing into oblivion, at any time there would most likely be only one communicative planetary system throughout the galaxy. We would have already met the Milky Way's only culture, for it would be us. One of the most compelling reasons to search for evidence of extraterrestrial civilizations, Morrison thought, would be to learn whether our own had a prayer of surviving its current technological adolescence. Maybe a message from the stars could provide some inoculation against humanity's self-destructive tendencies.

Sagan attempted to counter the doomsaying, noting that we could not rule out some technological civilizations achieving global stability and prosperity either before or even after developing weapons of mass destruction. They might master their planetary environment, and move on to exploit resources in the rest of their planetary system. He thought that such a society, flush with power and wisdom, would have a fighting chance to prevent or withstand nearly any natural calamity. It could, in theory, persist for geological timescales of hundreds of millions or even billions of years, potentially lasting as long as its host star continued to shine. And if, somehow, that civilization managed to escape its dying sun and colonize other planetary systems . . . well, perhaps then it would endure practically forever. Of all the attendees, Sagan was by far the most optimistic that technological civilizations could solve not only their many planetary problems, but also the manifold difficulties associated with interstellar travel. Somewhere out there, if not in our galaxy then in at least one of countless others, immortals passed their unending days amid the stars. Sagan thought we might yet be included in their number.

After the participants had discussed and debated L to the point of exhaustion, Drake stood up and announced that they had reached a consensus. The lifetimes of technological civilizations, he said, were likely to be either relatively short, lasting at most perhaps a thousand years, or very long, extending to one hundred million years and beyond. If indeed longevity was the most crucial consideration of the Drake equation, that implied there were somewhere between one thousand and one hundred million technological civilizations in the Milky Way. A thousand planetary civilizations translated to one per every hundred million stars in our galaxy. If the number was that low, we'd be hard-pressed to ever find anyone to talk to, as our nearest neighboring civilization would most likely be many thousands of light-years away. Conversely, if a hundred million civilizations existed, they would occupy one out of every thousand stars, in which case we might expect to have heard from them already. Drake's best guess in 1961 walked the line between these extremes: He speculated that L might be about ten thousand years, and that consequently perhaps ten thousand technological civilizations were scattered throughout the Milky Way along with our own. It was probably not coincidental that Drake's personal estimate rendered the successful detection of alien civilizations still quite difficult but not entirely beyond our capabilities: by his reckoning, only ten million stars would need to be monitored to obtain an eventual detection, though the search could take decades, even centuries.

At the conference's end, as the guests drank champagne left over from celebrating the news of Calvin's winning of a Nobel Prize, Struve offered up a toast: "To the value of L. May it prove to be a very large number."

CHAPTER 2

Drake's Orchids

A half century later, as we chatted in his living room, Drake expressed his conviction that most of the Green Bank conference's conclusions were, if anything, too pessimistic. In the last few decades the astrophysical case for a life-friendly universe had grown immensely, he said. Estimates of the rate of star formation had scarcely changed since 1961, but many new studies hinted that "red dwarfs," stars smaller, cooler, and far more plentiful than ones like our Sun, were more amenable to life than previously believed. Statistical analyses of data from the exoplanet boom suggested that hundreds of billions of planets existed in our galaxy alone, around all varieties of stars; the Green Bank group's original estimates of planet-bearing stars had been far too low.

Inevitably, a good fraction of all those planets would orbit within habitable regions of their systems. Spacecraft visiting Venus and Mars had pieced together tantalizing evidence for oceans of water on both worlds billions of years ago, though the planets' periods of habitability were brief, and after hundreds of millions of years each had lost its ocean. Meanwhile, researchers had discovered oceans of liquid water in the outer solar system, vast sunless seas beneath the icy crusts of gas giants' moons like Jupiter's Europa and Saturn's Titan. Extrapolating from these results, astronomers speculated that perhaps habitable Earth-like moons orbited some of the warm Jupiter-size worlds already known around other stars. A few even spoke of habitable planets free-floating through the depths of interstellar space after being slingshotted away from their stars. A thick atmospheric blanket of greenhouse gas or an icy crust over a deep ocean could insulate such nomadic worlds and preserve their habitability for billions of years. It could well be that most planets suitable for life in our galaxy don't orbit stars like our Sun, Drake said. Perhaps they didn't even orbit stars at all.

He thought the biochemical case had grown, too. A half century of progress in studying the origins of life had found a plethora of possible chemical pathways that could lead to membranes, self-replicating molecules, and other fundamental cellular structures. Multiple lines of evidence indicated that the jump from single-celled to multicellular life had occurred several times on the early Earth in a wide array of organisms, suggesting that the transition was yet another instance of convergent evolution, not a rare fluke. Researchers had discovered microbes flourishing in rock miles beneath the Earth's surface, in boiling-hot pools of hypersaline acidic water, in the icebox interiors of glaciers, in the deepest, darkest ocean abysses, and even in the radiation-riddled containment chambers of nuclear reactors. Once it arose, life as a planetary phenomenon appeared to be supremely adaptable, prospering in every possible ecological niche and enduring almost any conceivable environmental disruption.

I asked what all that meant for the later terms of his equation.

"We've found a truly great number of potentially habitable places, but the number of places where you could expect to find intelligent, technological life really hasn't increased that much," Drake replied. "That suggests to me there are probably significant barriers to the development of widespread, powerful technology. To surpass them, you might need a planet quite a lot like Earth. That may sound discouraging, until you realize just how many stars there are. Their sheer number suggests the equivalent of Earth and its life has probably happened many times before and will occur many, many times again. They're out there."

He chuckled, coughed, and creakily unfolded himself from the couch, clearly weary of sitting. We went outside to breathe fresh air.

Afternoon sunlight warmed our faces, and a cool breeze sighed through the towering redwoods to tousle Drake's silver hair. The air smelled of green, growing things. Drake pointed out the Moon's thin waxing crescent, faintly visible high in the cloudless sky. It was adjacent to the passing silver needle of a high-flying passenger jet. As we walked down into the yard, I gingerly stepped over the pale blue remnants of a robin's egg cracked open on the front steps, fallen from a nest in an overhanging tree. The tide was rolling in far below us, down past the forested hills and beachfront suburbs, and surfers rode big waves toward the shore of Monterey Bay.

The scene from Drake's front door encapsulated many of the essential facts of life on Earth. Fueled by raw sunlight, plants broke the chemical bonds of water and carbon dioxide, spinning together sugars and other hydrocarbons from the hydrogen and carbon and venting oxygen into the air. Sunlight scattering off all those airborne oxygen molecules made the sky appear blue. Animals breathed the oxygen and nourished their bodies with the hydrocarbons, utterly dependent upon these photosynthetic gifts from the plants. In death, plants and animals alike gave their Sun-spun carbon back to the Earth, where tremendous heat, pressure, and time transformed it into coal, oil, and natural gas. Mechanically extracted from the planet's crust and burned in engines, generators, and furnaces, that fossilized energy powered most of humanity's technological dominion over the globe. Built up and locked away for hundreds of millions of years, the carbon stockpile was gushing back into the planet's atmosphere in a geological instant.

Our experience at Monterey Bay was a product of our planet's physical characteristics—and the unlikely events that led to them. Earth's abnormally large Moon, which stabilizes our planet's axial tilt and bestows it with tides, was born when a Mars-size body collided with the proto-Earth early in our solar system's history. Another impactor, a six-mile-wide asteroid, struck the Earth 66 million years ago and sparked a global mass extinction, ending the age of dinosaurs. Humanity's small mammalian ancestors began their slow progress toward biospheric dominance, and the saurians that didn't die out gradually gave rise to birds. Billions of years before the dinosaurs, the life-giving liquid we recognize as Earth's ocean was mostly delivered by impactors, too, in a shower of water-rich asteroids and comets from the outer solar system. Earth's aquatic abundance, it is thought, lubricates the planet's fractured crustal plates and allows them to drift and slide in the geological process we call plate tectonics, a climate-regulating mechanism unique to our world out of all those in the solar system.

Turning away from the bay, Drake walked over to the center of his driveway, where the weathered stump of a giant redwood rose like a long-extinct volcano. He stooped and placed his hands upon the ancient wood. Years ago, he said, he had spread a thin layer of chalk on a section of the stump's surface, allowing the growth rings to be easily seen, and set his young children to the task of counting them as an informal science project. They counted more than 2,000, one for each year of the tree's life, which apparently began around the time of the birth of Jesus Christ.

"This tree saw the first light from the supernova that made the Crab Nebula, right about here," Drake said, touching a point midway between the stump's center and perimeter. Light from the supernova washed over the Earth in 1054, just as Western Europe was emerging from its Dark Ages. Sweeping his hand halfway farther out toward the perimeter, he brushed over the Age of Discovery, past rings recording the years when Europeans first explored and colonized the Americas. His hand kept moving until it slid from the stump's edge.

Over the course of the tree's 2,000-year existence, the Milky Way had fallen nearly five trillion miles closer to its nearest neighboring spiral galaxy, Andromeda, yet the distance between the two galaxies remained so great that a collision would not occur until perhaps 3 billion years in the future. In 2,000 years, the Sun had scarcely budged in its 250-million-year orbit about the galactic center, and, considering its life span of billions of years, hadn't aged a day. Since their formation 4.6 billion years ago, our Sun and its planets have made perhaps eighteen galactic orbits—our solar system is eighteen "galactic years" old. When it was seventeen, redwood trees did not yet exist on Earth. When it was sixteen, simple organisms were taking their first tentative excursions from the sea to colonize the land. In fact, fossil evidence testified that for about fifteen of its eighteen galactic years, our planet had played host to little more than unicellular microbes and multicellular bacterial colonies, and was utterly devoid of anything so complicated as grass, trees, or animals, let alone beings capable of solving differential equations, building rockets, painting landscapes, writing symphonies, or feeling love.

By its twenty-second galactic birthday, some thousand million years hence, our planet may well return to its former barren state. Astrophysical and climatological models suggest that by then the Sun, steadily brightening as it ages, should increase in luminosity by about 10 percent—a seemingly minor change, but enough to render Earth's climate too hot and its atmosphere too anemic to support complex multicellular life. Around that time, the oceans will begin evaporating, and most of Earth's water will rapidly cook off into space. The loss of oceans a billion years from now marks the most likely expiration date for all life on Earth's surface, though the omnipresent microbial biosphere might endure for billions of years further, shielded deep within the planet's parched crust. Somewhere in the neighborhood of five billion years from now, the Sun will exhaust its supply of hydrogen and begin fusing its more energy-rich helium, gradually ballooning 250 times its current size to become a red giant star. Astronomers debate whether the Earth will be submerged within the hot outer layers of the swollen red Sun or whether it will escape relatively unscathed and only suffer its crust being melted back to magma. Either way, at that late date the life of our planet will be brought to a decisive conclusion.

Considering the long concatenation of astrophysical events that led to our habitable planet, and the unknown synergies of technology and geology that could shape its fate, the distinction between chance and necessity blurs. Given a few hundred million years, would life arise on any rocky, wet, warm world? Would intelligence and technology emerge only on worlds with histories that mirrored our own, replete with the equivalents of Earth's Moon, mobile crust, and blue sky? Or was a focus on these features merely a failure of our Earth-bound imaginations? Was our planet and its history a useful template or a stumbling block in the search for alien life and intelligence? Would we even recognize our own planet as "Earth-like" if we glimpsed it a half billion years in its past or in its future? Answers to questions like these would be elusive as long as scientists only had one living world to study—our own. Drake didn't believe they would remain intractable forever.

Back in 1960, I thought that the possibility of detecting extrasolar planets in my lifetime was very, very low, though Otto Struve had already given us ideas about how it might someday be done," Drake had told me back in his living room. "I thought our only hope of detecting evidence of other planets was to receive radio signals from any intelligent creatures on them. We're seeing a similar pessimism play out now with characterization of planets around other stars. The techniques are there before us."

Already, planet hunters had found a handful of worlds that in their most basic details didn't appear too dissimilar from Earth. Those planets, their numbers growing every year, could potentially be much like our own. But the methods used to find them relied on closely observing a planet's bright, beacon-like star, not the dim planet itself; the gravitational pull of a planet on its star, or the shadow a planet cast toward Earth as it transited across its star's face, generally only revealed such things as a world's mass, size, and orbital properties. Without actually seeing these worlds—that is, collecting and analyzing photons reflected off their atmospheres and surfaces—scientists would be unable to determine whether any potentially habitable, potentially Earthlike planet was actually either of those things. They would be stuck where Drake had been fifty years before, hoping against all odds for a message from the stars to come streaming from the sky, filled with information on the flora, fauna, and environment of a place far, far away.

During the nineteenth century, a series of incremental discoveries led to the breakthrough that enabled the bulk of modern astronomy: light emitted, absorbed, or reflected by matter changes its colors in a way that captures the matter's chemical signature. Splitting up light into a spectrum to reveal those colors—a technique called spectroscopy—reveals those signatures, allowing astronomers to remotely measure the chemical composition of galaxies, stars, and planets. If they could somehow take a promising exoplanet's picture by gathering enough of its reflected photons, researchers could use the resulting spectrum to investigate that world's atmospheric chemistry. They could search for indicators of habitability, such as water vapor and carbon dioxide, as well as signs of life, like the free oxygen that filled and tinted our own planet's skies. They could look for the glint of a parent star's light shining off the smooth, flat surface of a planet's oceans or seas, or even subtle changes in the color of land that would hint at photosynthetic plants. Astronomers using observations from satellites and interplanetary spacecraft had already performed all these measurements for the Earth, confirming that our living planet could, in theory, be studied from across the vast distances of interstellar space.

Even if any extraterrestrials didn't advertise their presence to the universe at large, techniques like spectroscopy offered hope that we could still find and study their home worlds.

In the last decades of the twentieth century, as exoplanetology became a legitimate scientific field, planet hunters devised several ways to take planetary snapshots across the light-years. All involved one or more custom-built space telescopes designed to nullify a target star's glare and reveal its retinue of planets. At a likely cost of several billion dollars, a single space telescope could be built capable of delivering images of worlds around nearby stars, each planet manifesting as a dot a few pixels wide—minuscule, but more than enough for atmospheric spectroscopy. If money were no object, a fleet of telescopes could be assembled in space or on the far side of the Moon to act as one giant instrument, yielding larger images of nearby exoplanets that, though still very low-resolution, could reveal a world's shorelines, continents, and cloud patterns. Such telescopes would go a long way toward determining whether a planet was worthy of being anointed "Earth-like." Based on a fragmented astronomical community, an apathetic public, a gridlocked political system, and a struggling global economy, however, none appeared likely to be built anytime soon—at least not by the federal government of the United States of America.

Drake felt that if something could happen, somewhere it would happen, even if not right here and now. He wondered whether, if advanced cultures existed around nearby stars, they might have been watching our planet for quite some time using large space telescopes of their own.

"I'm speculating far out on a limb here," he said as we walked around his yard. "But I would guess that most every civilization with technological capabilities slightly beyond our own uses lenses on the order of a million kilometers in diameter to explore the universe and communicate between stars."

Beginning in the late 1980s, Drake had begun exploring an idea that made a lunar far side dotted with telescopes seem like child's play.

In retirement, the work had come to consume him, and now occupied much of his remaining time. He wanted to create a telescope that would surpass all others, one with a magnifying lens nearly a million and a half kilometers in diameter. Drake had found a way to transform the Sun itself into the ultimate telescope.

A consequence of the Sun's immense mass is that it acts as a star-size "gravitational lens," bending and amplifying light that grazes its surface. This effect, first measured during a solar eclipse in 1919 by the astronomer Arthur Eddington, was one of the key pieces of evidence that validated Einstein's theory of general relativity. Simple math and physics, judiciously applied, show that our star bends light into a narrow beam aligned with the center of the Sun and the center of any far-distant light source. As first calculated by the Stanford radio astronomer Von Eshleman in 1979, the beam comes into focus at a point beginning some 82 billion kilometers (51 billion miles) away from the Sun, nearly fourteen times farther out than the orbit of Pluto, and extends outward into infinity. There are as many focal points and Sun-magnified beams as there are luminous objects in the sky—imagine a great sphere surrounding our star, its surface painted with amplified, high-resolution projected images of the heavens.

Reviewing Eshleman's calculation, Drake had discovered that, due to electromagnetic interference generated by ionized gas in the Sun's outer layers, ideal seeing conditions for this ultimate telescope weren't at 82 billion kilometers, but almost twice as far out, at a distance of 150 billion kilometers (93 billion miles), a thousand times our distance from the Sun. For perspective, in June of 2011, humanity's fastest and most-distant emissary, the *Voyager 1* spacecraft launched in 1977, was just under 18 billion kilometers from the Sun, a bit more than a tenth of the distance to Drake's ideal focus. It had taken thirty-five years to get that far from Earth. Clearly, utilizing our solar system's ultimate telescope was a goal that could potentially take centuries to achieve. But the payoff might be worthwhile. Placed at any distant object's given focal point, a light-gathering telescope on the order of 10 meters (33 feet) in size could beam images back to Earth about a million times higher in resolution than what a network of large telescopes on the lunar far side could deliver. If, for instance, we wished to examine a potentially habitable planet orbiting one of the two Sun-like stars in Alpha Centauri, the Sun's nearest neighboring stellar system, a 10-meter telescope aligned with the Sun–Alpha Centauri gravitational focus could resolve surface features such as rivers, forests, and city lights. Put another way, a gravitational lens at Alpha Centauri could easily see the coastline of Monterey Bay, its tree-covered hills, and the bright lights of nearby big cities like San Francisco and Los Angeles.

"One of the beauties of gravitational lenses is that since the lensing object bends space, all light traveling through is equally affected," Drake said, squinting into the sunlight beneath one of his lemon trees. "Gravitational lenses are achromatic—they work the same for optical light, infrared, everything. I like to think of what they could do for radio. If you had two civilizations around different stars in communication and aware of each other, they could use gravitational lensing to set up transmission and receiving stations on each end. You look at the numbers, and at first it seems totally insane, but this is real. You could transmit, let's see, high-bandwidth signals from here to Alpha Centauri using only one watt of power. . . ."

He looked at me expectantly, but I could think of nothing to say.

"That's the transmitting power of a cell phone," he finished. "There's a quote I sometimes use when I talk about this, from a French play called *The Madwoman of Chaillot*: 'I know perfectly well that at this moment the whole universe is listening to us—and that every word we say echoes to the remotest star.' The capabilities of gravitational lenses make that sort of paranoia almost justified. If there's enough capability out there to build these things, you could have a kind of 'galactic internet,' with everyone monitoring and talking to each other, all with very high bandwidth and very low power."

• • •

After a half hour of outdoor ambling, we found ourselves standing before Drake's trio of greenhouses. They were where he spent much of his time when he wasn't caught up in his SETI work. He opened the door to the nearest one, and the hum of ventilation fans and a blast of humid, loamy air flowed out over the grass. Stepping inside, he let out a peaceful sigh. Like the other two greenhouses alongside it, this one was filled with orchids. Orchids hung from the translucent roof in pots of sphagnum moss, orchids stretched in rows on long wooden tables strewn with watering cans, and orchids sprouted from plastic buckets beneath lamps and irrigation tubes. Drake said he had about 225, but most were dormant. I counted only about a dozen blooms across the three greenhouses. He had picked up the hobby in the 1980s, about the same time he began seriously thinking about using the Sun as a gravitational lens. He did it for the challenge, he said, of nurturing the sometimes temperamental plants into full bloom, and for the satisfaction of seeing beautiful new morphological varieties emerge. Over millions of years, natural selection had shaped orchid flowers into a rich diversity of shape and color, each variety typically tuned to one or two species of pollinators. "Insects, mostly beetles," Drake said. "They blindly shape the flowers. But the hybrids, of course, are chosen and bred by humans."

Drake flipped on a grow lamp overhead, and in its pinkish light showed me a few blossoming hybrids, some cultivars he had cross-pollinated by hand. Each was wildly different from the others. One bore tiny flowers with trailing white petals and anthers heavy with yellow pollen. Another had five tubular, drooping purple blooms, each surrounded by a starburst of red-tinted curly leaves.

Drake turned to what he said was his current favorite, a single orange bloom with three angular petals that tapered to sharp, blood-red points. They looked like fangs. "This one's a hybrid of two different ge-nuses, *Dracula* and *Masdevallia*," he said. "Cold-growers from the Andes. No one's seen one like this before, with this red. It wasn't bloom-ing yesterday. Some of these only blossom one day out of the year, and the next day they're gone. You're lucky to be here right now—the flow-ers aren't long for this world." He touched the petals with reverence.

"They die, but they have reincarnation," Drake went on. "In prin-ciple, well-tended orchids are immortal. They reproduce by putting out new growths. Here's one." He gestured at a plant that bore no blooms but had several yellowish bulbous shoots hanging from its en-casing pot. "This one is quite old. It's outgrown its container—I should probably transfer it. You can see its new growth in these pseudobulb leads. Once you have two or three of these, you can cut one off and plant it in fertile soil. It becomes a new plant, and that plant will make more, and those plants will make more still. Each one doesn't live for-ever, it lives maybe three or four years, but the organism moves on like a wave, constantly generating new growth."

I told Drake his orchids made me think of L, a technological civi-lization's longevity, the greatest uncertainty in his equation. If it was too low, our galaxy could give birth to millions, even billions, of civili-zations over its eons-long life, but each one, isolated on a lonely planet, would wither and fall unseen with no chance for cross-pollination. If L was high, then in-bloom civilizations could linger and eventually inter-mingle, hybridizing their cultures across the light-years. Stability could set in; some would perhaps gain a sort of immortality.

Drake smiled and nodded. The similarity had not escaped his notice.

Back inside, Drake fished a bag of cashews from his cupboard and of-fered me a bottle of Sam Adams beer. He opened a can of Coca-Cola, and we sat down on his living room couch to discuss what the future might hold for SETI. Drake said he still thought that a civilization's average longevity approached 10,000 years, and that some 10,000 alien cultures were probably sitting out there in the Milky Way, waiting to be discovered. He admitted his belief was somewhat faith-based.

"I think 10,000 is plausible, but my estimate shouldn't be dignified by saying there's observational evidence that could accurately lead you to that specific number," he said between mouthfuls of cashews. "The factor of L still remains a total puzzle. We now know the rough fractions of stars with planets, and we're closing in on the frequencies of habitable planets. Sooner or later we'll know that number. But something like the lifetime of technological civilizations . . ." He trailed off, and stared for a long moment at the living room's blue stained-glass window.

Bits of multicolored glass were fused within the window's field of blue, forming a series of pictograms outlined in metal wire. Sunlight shining through gave the window a phosphorescent glow like an old analog television screen, and the colorful, blocky shapes looked very much like crude pixelated graphics from some lost, early-1980s video game. Drake had devised the design in 1974, when he was in the middle of a two-decade stint as a professor at Cornell University. Drake had initially been drawn to the job in 1964 because at the time Cornell managed the newly opened Arecibo Observatory, our planet's largest and most powerful single-aperture radio telescope. Soon after arriving at Cornell, Drake became the director of Arecibo, a position he held until 1981. The observatory was built into an immense limestone sinkhole in the jungle of northern Puerto Rico and boasted a 305-meter-wide (thousand-foot) bowl-shaped aluminum dish—big enough, Drake once calculated, to hold more than 350 million boxes of corn flakes. It was also big enough to transmit messages across hundreds, even thousands of light-years. On November 16, 1974, Drake used the massive dish to blast his message on a focused pencil beam of modulated radio waves toward a star cluster called M13, located some 25,000 light-years away, in the constellation of Hercules. With an effective radiance of twenty million megawatts at its specific wavelength, for the three-minute duration of the transmission Drake's beam outshone the Sun by a factor of 100,000.

The image's low resolution was a functional necessity; its content was formed from a series of 1,679 frequency pulses in the transmission beam: 1,679 is the product of two prime numbers, 73 and 23. Thoughtful aliens, Drake hoped, would use this hint to correctly interpret the message's pulses as forming a grid of 0's and 1's 73 units high and 23 wide. His stained-glass window displayed the resulting output: a top row of dots establishing a binary counting method, listing numbers one through ten, followed by a second row listing the atomic numbers of hydrogen, carbon, nitrogen, oxygen, and phosphorus, the key chemical elements of all life on Earth. A third section assembled the preceding atomic numbers into chemical formulas for the nucleotides in a molecule of DNA, followed by a schematic depiction of a DNA molecule's distinctive double helix. A long vertical bar represented the DNA molecule's sugar-phosphate backbone, and doubled as a binary depiction of 3 billion, roughly the number of nucleotide base pairs within the human genome. The molecule's image hovered over the head of a stick-figure human being, which was sandwiched between two more binary numbers, 4 billion and 14. Four billion was meant to convey the world population in 1974, and 14, multiplied by the transmission's wavelength of 12.6 centimeters, was intended to show that the human figure stands 176 centimeters high—just as tall, it turns out, as Frank Drake. The figure stood above the third of nine small dots extending out from one dot very much larger—a representation of our solar system and a hint that we lived on the third planet from our star. Finally, Arecibo itself was depicted as a simplified dish and antenna, with its gargantuan dimensions given in binary notation.

Whether any aliens would comprehend the message is another matter—even for most humans, it was largely indecipherable. When Drake showed it to his peers prior to transmission, he found that their grasp of its content varied widely based on their expertise. Chemists spotted the elements, astronomers discerned the solar system, while biologists and most everyone else recognized the DNA. But not a soul correctly interpreted each and every element of the Arecibo message.

In the years after Drake's Arecibo transmission, the question of its eventual interpretation was rendered somewhat moot by the realization that, a bit less than twenty-five millennia from now, when the message's photons should be reaching some 300,000 stars in M13, they will pass instead through empty space. Galactic rotation will have long since carried the star cluster far from the message's targeted swatch of sky. The pulse of radio waves will continue onward, perhaps passing near a few solitary suns before eventually escaping the confines of the Milky Way. Its fading echo of technology, its low-resolution snapshot of a biology, a culture, will stream on without end though the intergalactic void, long after Earth itself is but a memory.

The Arecibo message was more than the sum of its parts; it arguably represented the pinnacle of Drake's personal and professional interest in interstellar communication. Indeed, Arecibo Observatory was something of a linchpin for his dreams of messages from faraway lands and strange peoples. Over the years, as the lingering cosmic silence led some SETI practitioners to lower their estimates of L and of the likelihood of civilizations around nearby stars, the giant Arecibo dish became a central justification for continued SETI efforts. Thanks to its existence, the possibility of contact could be preserved in an atmosphere of increasing pessimism: even supposing that our nearest neighbors were halfway across the galaxy, if they possessed and transmitted with something like our own Arecibo, built with early-twentieth-century technology, we could in principle still detect their signals. Arecibo was, for instance, one major rationale for the Allen Telescope Array's later survey of planet-hosting stars in the Kepler field of view. Most of the Kepler field stars are several hundreds of light-years distant; the ATA would be extremely unlikely to detect a radio signal from any of them unless the transmitting dish at the other end was at least as large as the one at Arecibo.

Much like the SETI Institute and the ATA itself, Arecibo had seen better days. Around the turn of the millennium, its funding had begun a steady decline as federal agencies such as the National Science Foundation and NASA, struggling with politically driven budget cuts to their own bottom lines, slashed their spending on the Observatory. Private donations, university funding, and a modicum of financial support from the Puerto Rican government did not sufficiently increase to fully offset the resulting shortfalls. In May of 2011, the NSF announced that Cornell University would no longer manage the Observatory, and handed off the responsibility to a consortium of public and private management partners, led by the nonprofit organization SRI International. Rumors circulated that, in the event of no major additional funding sources being found, the new managers would shut down Arecibo Observatory, dismantle the world's largest radio dish, and return the limestone sinkhole to its natural state. In 2012, SRI International was given stewardship over the ailing ATA as well.

"In the beginning, the L factor was simply the likely duration that a civilization possessed high technology," Drake said, turning his attention away from the window and back to the bag's dwindling supply of cashews. "L really should be the length of time that a civilization has technology that you, that we, can detect. And that muddles everything up, because it means L depends not only on there being a technology to detect in the first place but also on the technological capabilities of the searchers. Look at our own civilization, for example. We've had radio for, so far, about one hundred years, which you'd think would give us a minimum L of one hundred. But we're now becoming far more radio-quiet, so if someone's looking at us with radio, they might not see us much longer.

"Back in the 1960s, we had powerful military radars, early-warning systems against intercontinental ballistic missiles, things like that," Drake reminisced. "Those could be detected from nearby stars using equipment similar to what we had back then. I thought at the time that sort of technology would just keep getting more powerful, and that would keep Earth visible practically forever. What actually happened is that technology did get more powerful, but not how I'd expected. It got more efficient. The switchover to digital television has made us much, much less detectable than when we used analog broadcasting, for instance. We send more through coaxial cable and optic fiber than we used to. And most of the ways we transmit radio signals now are almost indistinguishable from cosmic radio noise. All that causes what could have been a big sign of our existence to just vanish. Poof!"

Drake sighed. "These days I think that more-advanced technological civilizations will probably prove more difficult to detect than younger ones," he said.

On Earth, the high technology developed during the first half of the twentieth century had in the second half spread from developed countries to colonize the entire globe. After harnessing the power of the atomic nucleus, science had turned to the machinery within the nuclei of living cells, bringing forth what promised to be a transformative era of synthetic biology. The world's human population had more than doubled, driven by bioengineered boosts in agricultural productivity, breakthroughs in medicine, and a host of other science-fueled increases in living standards. Simultaneously, extinction rates of natural species had soared due to environmental disruption and habitat destruction. The land was laced with superhighways, power transmission lines, and fiber-optic communications networks; the sky was crisscrossed with transcontinental jet contrails and the starlike gleams of orbital satellites; the air itself was filled with electromagnetic chatter from radios, televisions, and mobile phones, as well as with rising amounts of carbon dioxide from the frenzied combustion of the planet's reservoirs of fossil fuel. Rapid, successive revolutions in information technology had made powerful computers networked, ubiquitous, and personal, creating vast virtual realms often only tenuously linked to the world of atoms.

What those changes meant for the future of our culture and our world remained to be seen, though it seemed possible that, given a few centuries' time, we might not even recognize whatever our descendants had become. I mentioned to Drake that many of the same Silicon Valley tycoons who helped fund the SETI Institute often chattered about a dawning era of even more radical and rapid change, a coming "technological singularity" in which exponential growth in computing power and sophistication would profoundly transform, at minimum, the entire planet. Some techno-prophets spoke worshipfully or fearfully of computers becoming sentient and gaining godlike powers. Others speculated that someday humans would break free of their carbon-based chains by uploading their minds into silicon substrates, where they could, in some manner, live forever. All seemed to agree that if humans themselves weren't destined to inherit the Earth, they would certainly author whatever ultimately would. A few even conjured up the bygone Space Age dreams of Drake's youth, envisioning a new golden era of prosperity and exploration in which humans would travel with their intelligent machines throughout the solar system, and perhaps someday to other stars.

"Yeah, I've heard all that stuff," Drake replied. "It would be nice if we made it to Mars. But I don't hold with the hypothesis that we'll all slowly become or be replaced by computers. And of all the things we might someday do, I don't think we'll ever colonize other stars."

I asked why not.

"I don't think computers can have fun," he said. "I think joy is a quality not available to computers. But what do I know?" He laughed. "Interstellar travel, on the other hand, I've worked on that quite a bit. Putting a hundred humans around a nearby star costs about a million times as much as putting them in orbit in your own system. You'd have to be pretty rich to pull that off.

"Let's say you have two colonies ten light-years apart—that's probably the typical distance between habitable planets, I'd guess. The fact is, you can't really go faster than about a tenth of light-speed. At speeds higher than that, if you hit anything of any substance whatsoever, the amount of energy released approaches that of a nuclear bomb. So you're limited to about ten percent, a speed we currently can't come anywhere close to, and that means you're looking at journey times of at least a hundred years. The distances, times, and speeds are daunting, but the most daunting thing of all is the cost. Take something the size of a Boeing 737 plane, which is about the smallest that might make a reasonable crewed expedition, and send it at a tenth the speed of light to a nearby star, okay? Now just work out the kinetic energy that's in it. It turns out to be about equal to two hundred years of the total electric power production in today's United States. And that's assuming a one-way trip, where you don't even slow down and enter orbit on the other end. The inherent difficulty of interstellar travel is one of the big reasons why looking for things like radio signals is so appealing."

"So you think we're stuck in the solar system," I said, thinking of distant days when the swollen red Sun would sterilize Earth. "This is it?"

"Yeah, I think so," Drake somberly replied. "You have to admit, though, that it's pretty good while it lasts."

Drake ate the last of his cashews, picked up his can of Coca-Cola, and tilted its lip to clink against the neck of my beer bottle. We drank to *L* and to all those who sought to make it a larger number.

CHAPTER 3

A Fractured Empire

When Project Ozma was unveiled in 1960, it created a deep rift among astronomers. Some loved the idea of scouring the skies for other galactic civilizations, while others thought it the worst form of pseudoscience. In SETI's defense, Otto Struve drafted an influential letter circulated among the upper echelons of the global astronomy community.

In the letter, Struve emphasized that planets were probably common around other stars, and that, while the likelihood of life or intelligence emerging on any particular world was unknown, "an intrinsically improbable single event may become highly probable if the number of events is very great. . . . There is every reason to believe that the Ozma experiment will ultimately yield positive results when the accessible sample of solar-type stars is sufficiently large." Humanity, he reasoned, could no longer consider itself alone and anonymous on the cosmic stage.

Astronomy was at a turning point, Struve wrote. The Space Age had thrust the field into "a state of turbulence, uncertainty, and chaotic expansion unknown in the history of mankind," one increasingly funded by massive government coffers. Astronomers could capitalize on that newfound abundance by embracing the search for extraterrestrial life and intelligence, mustering a new age of discovery rivaling that of the Enlightenment. Or, they could just muddle along pursuing nothing but the status quo, leaving a less impressive record for future historians, one defined "by the team work of many competent but not especially brilliant scientists, by the evident confusion of ideas, by the competitive aspects of our research and its political overtones." The truth, as it so often turns out, would lie somewhere in between.

Struve's admonishment was on my mind when, a few days before my meeting with Drake, I attended a gathering of scientists and journalists 125 miles north of Santa Cruz, at the Marconi Conference Center in Tomales Bay. Built in 1913 by the Italian radio pioneer Guglielmo Marconi, the Center had in a previous life been the world's first trans-Pacific receiving station, though it now served as the site of an annual interdisciplinary conference held by the University of California, Berkeley's Miller Institute for Basic Research in Science. It was a warm sunny Saturday afternoon, with small boats and Jet Skis dotting the narrow bay's emerald waters, but the conference-goers were all sitting in a stuffy, darkened room, spellbound by a tall, smartly dressed man, thin and angular, with dark hair, wide green eyes, and a hawkish nose. He was talking excitedly, occasionally stammering in his rushing words, making gangling gesticulations in front of PowerPoint slides on a projector. He was Greg Laughlin, the forty-four-year-old astrophysicist and professor at UC Santa Cruz.

"I took this picture from my front door using an off-the-shelf five-megapixel camera," Laughlin said, pointing to what looked like a lump of pale Legos against a deep blue background. "That's Venus, zoomed in so that the individual pixels are visible. It's emblematic of the situation we're now in with exoplanets, in that we can see there's some structure there, it's mysterious, and we want to know more. It's also emblematic in that most of the things we're trying to understand about worlds orbiting other stars, we've already been through with planets in our own solar system."

Venus was also symbolic of Laughlin's early scientific beginnings. His first brush with astronomy had come when, as an eight-year-old boy in the soybean country of downstate Illinois, he scraped together enough money to buy a small, simple telescope. He looked at stars, and the Moon. In one early evening's twilight he turned his telescope toward Venus, low and sparkling in the sky. He had expected to see the same blurry dot he witnessed with his naked eyes, albeit magnified. Instead the telescope revealed a crescent, sharp and whitish-yellow like a nail clipping. It dawned on him that he was seeing both day and night on Venus, and that the demarcation between the two marked a zone of twilight, just like the one he was now in on Earth. The view from his backyard in Illinois seemed at once larger and smaller than ever before; something about seeing an alien planet's hidden details revealed before his very eyes made his mind effervesce. The feeling faded, only to momentarily return over the years each time he uncovered something unexpected and beautiful. The more he learned, the more profundity he saw in the purity of numbers and equations, the more majesty he found in the lives of planets, stars, and galaxies. Laughlin didn't know it at the time, staring at sunlight shining on the distant cloud tops of Venus, but the vista spread before him in his telescope would draw him deep into the frontiers of planet hunting.

"Though closer to the Sun, Venus is covered by so many clouds that it actually absorbs less sunlight than Earth does," Laughlin was saying to his audience. "And so for many years it was perfectly possible to imagine the Venusian surface looking like this." On the screen behind him, an aerial photograph appeared of waterfalls cascading through a mountain-ous forest shrouded in mist. "Then, in the late 1950s, astronomers found that Venus was just spewing out microwaves with an emission corre-sponding to a temperature on the order of six hundred degrees Celsius. It soon became clear that Venus was a runaway-greenhouse world, a truly terrible place." Laughlin summoned an image of the true Venusian surface and let it linger silently on the screen—a lifeless, flattened land-scape of shattered rock beamed back in 1982 by the Russian *Venera 13* lander moments before the probe melted and imploded beneath hellish temperatures and crushing atmospheric pressures.

"During the 1950s, there was this brief, wild interval when you could realistically speculate that Venus not only had a habitable envi-ronment, but also that humans would soon visit it. The Apollo pro-gram wasn't far away, and the ability to travel between planets was just within our grasp—in a way that it doesn't seem to be any longer. Think what would have happened, how history would have changed, and what our world would look like today if we had found a habitable Earth-like planet literally right next door. It's an odd, tragic coincidence that these possibilities disappeared for us right around the dawn of the Space Age. And as soon as Venus and then Mars changed from being candidates for full-blown economic colonization to being objects of mostly scientific interest, public interest really shifted to planets orbit-ing other stars."

A few slides later, Laughlin showed a graph plotting all the known exoplanets, with masses on the vertical y-axis and years of discovery on the horizontal x-axis. A lone dot resided in 1995's column on the plot's older, sparsely populated left side, high up between the masses of Jupi-ter and Saturn. The dot represented a gas-giant world in a star-grazing 4.5-day orbit around the nearby star 51 Pegasi. It was 51 Pegasi b, a "hot Jupiter," the first confirmed exoplanet around a Sun-like star, and a planetary system so bizarre that it spurred theorists to rewrite their mod-els of planet formation. Sweeping forward in time, the dots proliferated across the chart and spread out into a thick wedge spanning a wide range of masses. In a decade's span, the number of confirmed worlds beyond the solar system had soared into the hundreds, with no obvious end in sight. The field of exoplanetology was booming as never before.

Nearly all those worlds had been detected through a technique called radial-velocity (RV) spectroscopy, which watched for stars that wobbled from the gravitational to-and-fro tug of orbiting planets. When a planet tugs its star toward an observer on Earth, the waves of light from the star are compressed toward the blue end of the spectrum; when a planet tugs its star away, the starlight is stretched out toward the red. The same effect can be distinguished in waves of sound rather than light, as when an ambulance's siren rises in pitch as it roars toward you up a street, then falls in pitch as it speeds away. The frequency of a star's wobble indicates a planet's orbital period—its year. The wobble's strength—whether it corresponds to a kilometer or a centimeter of motion per second, for instance—yields an estimate of a planet's mass.

It is not easy to distill planetary signals from the motions of a million-kilometer-wide roiling ball of plasma we call a star, particularly when the planets are small and in more-distant orbits. To do so requires not only large telescopes, but also high-resolution ultra-stable spectrometers. A telescope's mirror gathers and amplifies light from a target star, which is sent streaming into such a spectrometer. Within the spectrometer, the starlight passes through a labyrinth of mirrors, gratings, and prisms that shape, split, and sort the photons by their wavelengths before sending them to be captured and preserved in the memory of a charge-coupled device, a CCD akin to those in consumer digital cameras. A star's raw spectrum looks like a stretched-out, chopped-up rainbow, its languorous red-to-blue continuum broken by thousands of black absorption lines. The lines come from particular atoms and molecules percolating at the star's glowing surface, soaking up certain wavelengths of starlight before the photons can escape to space. A star's wobble is discerned in these lines as they migrate redward, then blueward, in time with the star's reflexive motion from an orbiting planet's gravitational pull. To track the motions of the lines, astronomers project reference marks onto the spectrum like tiny ticks on a ruler. For small planets, the offset in a line's position may only be a fraction of a single pixel in the CCD detector—cooling the detector to cryogenic temperatures helps minimize stray electronic noise in the pixels, allowing such faint gradations to be seen. Stray electrical currents or minor changes in air pressure and temperature can also create noise that masks or mimics planetary signals. Complex statistical reductions and analyses of all the gathered data introduce further opportunities for error. Teasing faint RV signals from the noise is part straightforward science, part arcane art, and anyone who has the physical resources and mental capacity to do it belongs to an exclusive club with at most a few dozen members worldwide.

The problems of instrumental stability and data calibration were not new to planet hunting. Indeed, they had been behind an earlier, nearly forgotten era of false-alarm planets. Beginning in the early 1940s and extending into the early 1970s, several astronomers claimed detections of worlds orbiting nearby stars, all of which would ultimately prove illusory. The most prominent planet hunter of that bygone era was Peter van de Kamp, a Dutch-American astronomer at Swarthmore College. In photographic plates taken over a period of decades at the college's 24-inch Sproul Telescope, van de Kamp thought he spied planetary wobbles in the motions of Barnard's Star, a dim red dwarf and the next nearest star to our Sun after Alpha Centauri. His claims of two gas-giant planets around Barnard's Star were initially reported—and endorsed—in scientific journals and in such popular publications as *Time* magazine and the *New York Times*, but competitors could find no evidence of those worlds in their own observations. Subsequent investigations suggested that van de Kamp's wobbles had been due to aberrations caused by periodic cleaning and upgrades of the Sproul Telescope, as well as to faulty analytic techniques. Years of close surveillance never found further evidence of his planets. Van de Kamp died in 1995, a few months before the detection of 51 Pegasi b, never having forgiven his critics and unwavering in his certainty that his worlds were genuine. Astronomers today use his story as a powerful admonition against cavalier claims of planetary discoveries based on indirect evidence and weak statistics.

With the 2009 launch of NASA's $600 million Kepler Space Telescope, another more direct technique besides RV had become popular. Instead of looking for stellar wobbles, Kepler looked for transits, when a planet crosses the face of its star and blocks a small portion of the star's light as recorded on a CCD. The frequency of a transit's recurrence yields the transiting planet's year, and astronomers can estimate the transiting planet's size based on how much it dims a star's light. Unlike RV, which could over time conceivably detect most planets around most stars, transits rely on random geometric alignments. Only planets with orbits that, by chance, align approximately edge-on with the line of sight from Earth will transit. This means the vast majority of exoplanets are invisible to the technique. The gamble, however, was worth the jackpot. Surveying a single patch of sky containing more than 165,000 stars, at the time of Laughlin's talk the Kepler mission had already found in excess of 1,200 candidate transiting planets. "Candidate" was used until each planet could be confirmed or validated by other techniques, though many of Kepler's stars were too faint for robust follow-up measurements. By early 2013, the Kepler team had announced the discovery of more than a hundred confirmed worlds, and nearly 3,000 candidates.

Compressed and plotted onto the rightmost edge of Laughlin's chart, the announced Kepler candidate planets formed an unbroken line of dots. In comparison to the relative sparseness of all prior years, the Kepler data looked like a solid wall, one extending from several times the mass of Jupiter down to a few hundredths of Jupiter's mass— the mass of Earth. "Obviously, what this shows is that we're finding more and more planets at lower and lower masses," Laughlin said, gesturing at Kepler's wall of worlds on his chart. "Only a few years ago, almost all of this was terra incognita. Just this year, right now, we have finally begun detecting planets of one Earth mass. We've reached the point where we can actually credibly talk about planets around other stars that are the same size and mass as Earth, and we're starting to get a much better sense of how most planetary systems are arranged. We're taking what I like to call the 'galactic planetary census,' and what we've found is reminding us again that, like with Venus and 51 Pegasi b, making simple extrapolations based on the Earth or our solar system can be dangerous."

Laughlin played an integral part in the ongoing galactic planetary census, not as an observer at the telescope, but in analyzing the data the observers delivered. One of his specialties was the trial-and-error statistical process of piecing together a star's planetary system solely from RV wobbles. If a star's wobble was due to a lone orbiting planet, plotting the back-and-forth oscillation over time would yield a classic sine wave, with smooth and regular crests and troughs that repeated with the period of the planet's orbit, looking like the single pure note of a vigorously plucked violin string. Spotting that pattern in the data was simple. In multiplanet systems, however, each world imparts its own subtle distinct pull to the star, creating a more complicated pattern of wobbles. Teasing apart the system's architecture from those wobbles was rather like determining the layout and composition of an entire orchestra as each of its instruments played different notes all at once. If a planet hunter was too focused on a handful of isolated sweet tones in the music of the spheres, he or she could miss other planets hiding in the sour notes and residual noise. The smaller the world, the weaker its signal, and the more astronomers would strain to detect its presence in washes of stellar static. Laughlin had guided the development of a piece of software, the Systemic Console, that could discern planetary signals out of such complicated data sets. It had rapidly become one of the field's standard tools. Pulling up Systemic on his laptop, he gave his audience a taste of what real planet hunting was like. A black-and-white grid popped up onscreen, and hundreds of dots blossomed all over it.

"This is RV data for 61 Virginis, a star about twenty-eight light-years away, provided by my colleagues Paul Butler and Steve Vogt from two telescopes over the past several years," Laughlin explained. He clicked a button, and a sine wave superimposed itself over the data, intersecting many but not all of the dots. "Here's what would be the signal of a planet with a period of several hundred days and roughly a quarter of Jupiter's mass, but you can see it's nowhere near a perfect fit." He tweaked the model planet's orbit and mass for a few moments, but its sine wave stubbornly resisted aligning with the data points. "Now we'll just run the automated fitting routine, which tries out different stable planetary configurations, runs through optimizations, and comes up with its best fit." Another click of a button, and within seconds three distinct curves materialized out of the dots, running through far more of them than the sine waves from Laughlin's previous attempts.

"You can see the software's solution was three planets, although there are still residuals here—61 Virginis could easily have more planets, even in its habitable zone," he said. "This is a result my collaborators and I have published. What's really interesting is that out of the nearest few hundred stars in our local neighborhood, 61 Virginis is one of the most similar to our own. It has almost the exact same mass, radius, and chemical composition as the Sun, and it's of a similar age, and yet its planetary system is completely different. The masses of these planets, nearest to farthest, are something like 5, 18, and 23 Earth masses, and they're all closely packed roughly within what, in our own solar system, would be the orbit of Mercury. Our solar system has nothing interior of Mercury's orbit, but 61 Virginis has three whole planets packed down there! These sorts of architectures are turning up everywhere we look, yet they could not have been extrapolated from our own solar system. They were unexpected."

From the audience, someone asked how many stars might lack planets entirely.

"It's very hard to show that any star is devoid of planets," Laughlin replied. "So it's more useful to ask what percentage of stars have

55

planetary systems like ours. Data from RV surveys and from Kepler is now starting to show that stars with a Jupiter-mass planet in something like a ten-year orbit like our own system are fairly rare. This is something that I think at most only ten percent of surveyed stars can now support. At least as far as Jupiter is concerned, our solar system is somewhat unusual. Right now, the data are telling us that the archetypal planetary system is a Neptune-like planet in a warm, short-period orbit, but part of that is selection bias—those planets are easier to detect."

He began to deliberately pace the floor, regaining the rhythm of his presentation by walking back and forth between the podium and a window. "We really don't know much yet about the distribution of Earth-size planets in Earth-like orbits, but the expectation is that they will be abundant. Kepler's going to tell us soon, I think—it's easier to validate transits, even though they reveal small fractions of total populations. The small planets we're detecting today create RV signals on their stars of the order of a meter per second. I'm walking a meter per second right now. Now, that's an incredible accomplishment to detect such a small change in the motion of an entire star many light-years away, but it's not quite enough: the Earth's RV signal on the Sun is only about ten centimeters per second. Stars move more than that from interior pulsations and vibrations, and from material flowing around their surfaces—at any particular moment, a star is creating all that noise, introducing astrophysical jitter that contaminates the signal."

There was an implicit warning within Laughlin's carefully phrased remarks. The most tantalizing worlds—those that might resemble Earth and harbor life—were also some of the most difficult to find. A low-mass planet in a clement orbit about its star could often only be detected as a sliver of signal cresting above a sea of stellar noise. As the pressure mounted for RV surveys to turn up alien Earths orbiting nearby stars, it was also getting tougher to know what was actually real.

Faint RV planetary signals could be amplified, Laughlin told his audience, by lavishing a quiet star with attention, hammering away with hundreds or even many thousands of observations, all averaged over time to beat down the star's already low stellar noise. But the approach came with risks—a planet-hunting team fortunate enough to secure observing time on a world-class telescope and spectrometer might chase a star's tantalizing signals for months or even years, only to ultimately discover that its potential planets were illusory. Careers and reputations would be forged or shattered on the barest probabilistic whiffs of planets emerging from a statistical haze. "[The] push toward lower-mass planets is part of an 'arms race' among the different competing groups," Laughlin explained near his presentation's conclusion. "You find planets, and then the time-allocation committees give you more time to find more planets. If at any point you don't find planets, that's it, you're out of the game."

For more than a decade after astronomers began regularly detecting exoplanets in the mid-1990s, the RV "game" had been limited to a contest between two great planet-hunting dynasties, one American, the other European. The first began in Pasadena, California, in 1983, while a struggling twenty-eight-year-old astronomy postdoc named Geoff Marcy was taking a long morning shower. Marcy's research into stellar magnetic fields wasn't panning out, and had been roundly criticized by a few senior astronomers. He felt incompetent, depressed, and burnt-out. As the water streamed over his downturned head, he realized that up until then his career had been mostly a failure, and that if nothing changed, it might end before it had even properly begun. He thought back to what had set him down his disastrous path of astronomy, when he had been a young boy wondering whether all the stars in the sky had planets. Where had his passion gone? Suddenly, an epiphany: if he was destined for failure, he should fail spectacularly, pursuing a topic he loved but few others took seriously. By the time he finished showering, he had decided to spend the rest of his perhaps-brief career searching for exoplanets.

Marcy was not as incompetent as he believed. His encyclopedic knowledge of astronomy, paired with a quick wit and the skills of a natural storyteller, made even the most abstruse astronomical topics comprehensible to laymen. He soon landed a junior faculty job at San Francisco State University, and in between teaching courses he pondered an RV planet survey, though his plans always seemed half-baked—the spectroscopic signals of planets would be impossible to discern without proper calibration. Things coalesced when he met Paul Butler, a younger student simultaneously pursuing a bachelor's degree in chemistry and a master's in astrophysics. Butler shared Marcy's interest in exoplanets, and they became close friends. Together, they worked to find ideal calibration methods, until Butler came up with a solution: a glass vessel filled with iodine that could be attached to a spectrometer. Light shining through this "absorption cell" would project iodine absorption lines like hashmarks upon a star's spectrum, allowing small spectral wobbles to be seen. Butler's iodine cell would become the standard calibration technique for decades of RV planet searches.

In 1987, Marcy and Butler mated the iodine cell with the general-purpose "Hamilton" spectrometer built by Marcy's former PhD advisor, the UC Santa Cruz astronomer Steve Vogt, and began their planet search. For years they used the spectrometer on various telescopes at Lick Observatory on Mount Hamilton, twenty-five miles east of San Jose, searching to no avail for extrasolar Jupiters around 120 nearby Sun-like stars. Butler left for a time to obtain his PhD at the University of Maryland, but continued to hone the duo's data-analysis software, eventually sharpening the RV precision of their data from 15 to 5 meters per second. By the autumn of 1995 they were nearing the end of their patience when two University of Geneva astronomers, Michel Mayor and Didier Queloz, announced the discovery of 51 Pegasi b based on another RV survey conducted from the Haute-Provence Observatory in the south of France.

When they heard the news, Marcy and Butler rushed to observe 51 Pegasi for themselves, and within days saw the star's telltale wobble from its whirling hot Jupiter—a variety of planet they had not conceived of or looked for in all their previous years of searching. Revisiting their old data, they rapidly found two more giant planets around the stars 47 Ursae Majoris and 70 Virginis, retaking the lead in the burgeoning race of discovery and establishing a rivalry that would span decades.

In those early golden years, Marcy and Butler surged ahead in the race, propelled by nearly a decade of experience and their extensive back-catalog of data. By the turn of the millennium, they had discovered nearly forty close-orbiting gas giants. Each announcement was a news event—the discovery of exoplanets had yet to become truly routine. With their research featured on magazine covers and national newscasts, the duo abruptly found themselves in high academic demand, and soon secured more-prestigious positions. Marcy became a UC Berkeley professor, and Butler secured a job as a staff scientist at the Carnegie Institution for Science in Washington, DC. Though geographically separated, they continued their work together, ultimately utilizing their growing fame to expand their team to include Vogt as well as another brilliant planet hunter, the astronomer Debra Fischer. The group gained more funding and access to some of the best astronomy resources in the world, notably another Vogt-built spectrometer, HIRES, operating on the twin 10-meter telescopes of the W. M. Keck Observatory in Mauna Kea, Hawaii. HIRES could reach RV precisions of 3 meters per second, allowing it to discover smaller exoplanets in cooler orbits. But to reach potentially habitable worlds, even more precision would be required. Marcy began closing his intergroup e-mails with the exhortation "OMPSOD!"—One Meter Per Second, Or Death!

"The Swiss," as Marcy's and Butler's Geneva-based competitors are invariably called despite having collaborators around the world, were not sitting idle while their American counterparts surged. They were expanding their team as well, and redoubling their efforts to find more planets. As both teams excelled, their competition became fierce. At a conference in June 1998, the American team announced their discovery of Gliese 876b—the first planet found around a red dwarf star. The following day, the Swiss made an announcement of their own, saying they had detected the same planet days before the American announcement; they claimed to have confirmed their discovery some hours before the conference, but Marcy and Butler had beaten them to the podium. The American team would also beat them to peer-reviewed publication on the planet, and came away with credit for the discovery. In November of 1999, both teams nipped at each other's heels to share the discovery of the first transiting planet, a hot Jupiter around the star HD 209458, after separately and nearly simultaneously observing the transiting world and submitting papers on the results. The rivalry deepened in 2002, when the Swiss released a paper claiming the detection of twin "hot Saturns" around the star HD 83443. Marcy, Butler, and Vogt also observed the star, but could find no evidence in their data to support both Swiss planets. Butler spearheaded the publication of a paper detailing the case against the Swiss claims, and months afterward the Swiss team retracted one of the planets, placing a dark blemish on their otherwise flawless record. In contrast, while working together the American team never retracted a single world from their tally. The Swiss never forgot about the time when Butler led the American charge against them. They would keep him far from their orbit ever after.

The Swiss, for their part, seized the lead in RV precision in 2004 with the debut of their HARPS spectrometer, developed in collaboration with the European Southern Observatory (ESO). Stabilized in a temperature-controlled vacuum chamber and mounted on a 3.6-meter ESO telescope in Cerro Paranal, Chile, HARPS proved capable of breathtaking RV precisions of slightly below 1 meter per second, giving the Swiss a decisive edge that they used to discover a multitude of smaller planets orbiting on the cusps of habitable zones. Some of the worlds were only a few times the mass of our own, and thus possibly rocky rather than smothered in heavy layers of gas. They were hopefully called "super-Earths." Keck's HIRES would receive upgraded detectors that same year, boosting its precision closer to but not equaling that of HARPS. The Americans had more planets to their names, but they knew it was their competitors who were making faster progress toward detecting potentially habitable worlds. The Swiss had been the first to break the 1-meter-per-second barrier, and even with its upgrades HIRES was slightly below HARPS in performance. After their years of dominance, the Americans privately worried that their sudden disadvantages, albeit minor, could lead to the downfall of their team.

As the Swiss were developing HARPS in 2002 and 2003, Fischer, Vogt, Butler, and Marcy made grand plans of their own for what they hoped would be a superior piece of kit: the Automated Planet Finder, a 2.4-meter robotic telescope to be constructed at Lick Observatory and outfitted with a new spectrometer custom-built by Vogt to excel at precision RV measurements of a meter per second or less. Though it would be dwarfed by the light-gathering power of many larger ground-based telescopes, the APF's advantage would be its singular focus. Almost all world-class telescopes were by necessity workhorses for the entire breadth of astronomy, with only a portion of their time devoted to planet hunting. The APF's sole task, by contrast, would be to survey bright nearby stars, night after night, steadily accumulating RV signals for any accompanying small, rocky planets. The group selected Vogt to be the APF's principal investigator. The project hit a snag, however, after the relationship between Marcy and Butler took a sudden turn for the worse.

Over time, the duo's extreme success had pushed them apart and destabilized their friendship. They were no longer young—each man now sported dark circles beneath the eyes, a salt-and-pepper beard, and a head unburdened by more than a waning crescent of hair. After two decades of working together, much that had once seemed fresh and new now felt tiresome and constraining. Though he had initially been only a graduate student beneath Marcy, Butler's development of the iodine cell and seminal contributions to RV data-analysis techniques had elevated him to equal stature with his longtime research partner. Yet

Marcy still acted as the group's de facto leader and manager. Where Butler was taciturn and blunt, preferring the simplicity of actual planet hunting to the delicate nuance of press interviews and academic politicking, Marcy was loquacious, charismatic, and cunning, happy to speak at length in eminently quotable sentences about the team's work and always careful to offer diplomatic praise for his competitors. With the Swiss, Marcy was cordial, even friendly, though they had for years collectively treated Butler as persona non grata. The almost universal propensity to overlook Butler in favor of Marcy had progressively led to a tangible divergence of the two men's fortunes, as Marcy accrued the lucrative lion's share of press mentions and professional awards.

Feeling underappreciated and eclipsed, Butler reached his limit in 2007 and abandoned Marcy to form a new planet-hunting team with Vogt, using instruments at Lick Observatory as well as facilities in Chile and Australia. Their dynasty was splitting into multiple shards. Fischer soon left the team, and her position as a professor at San Francisco State University, for a professorship at Yale, where she founded a planet-hunting group of her own and began building a new spectrometer, CHIRON, meant to rival and surpass HARPS. Marcy, by now a coinvestigator on NASA's Kepler mission, remained at UC Berkeley and worked closely with a new protégé, the astronomer Andrew Howard, using Keck and HIRES to search for more planets and to study Kepler's thousands of candidates. Where once there had been only two RV teams in serious contention for finding alien Earths, the disintegration of the powerful Marcy-Butler partnership had given birth to many, with still more upstart groups waiting in the wings to use a new generation of planet-hunting spectrometers being constructed and deployed at observatories across the globe. But along with the unified Swiss, the scattered, grizzled veterans of the crumbled American dynasty still had the best data, the best observatory access, and the best chance of finding rocky planets in stellar habitable zones. Marcy's group, continuing its existing program while also first in line to drink from Kepler's firehose, was the odds-on favorite to be first.

On September 29, 2010, Vogt, Butler, and four other collaborators declared they had beaten those odds. Their announcement coincided, perhaps not by chance, with Marcy's fifty-sixth birthday—a final spiteful gift from former friends. By combining their own old HIRES observations with publicly available HARPS data from the Swiss, Vogt and Butler claimed the RV detection of two planets around the red dwarf star Gliese 581, located some 20 light-years away in the constellation of Libra. One of the planets was between three and four Earth masses, in a 37-day orbit that placed it squarely in the center of the habitable zone. According to convention, its official name was Gliese 581g—the "g" denoted that the world was the sixth planet discovered around that particular star, though it could have equally well stood for "Goldilocks," since the planet was potentially rocky, and in an orbit very clearly neither too hot nor too cold for life as we know it. Vogt preferred another appellation. He called it "Zarmina's World," after his wife, and said in a press conference that he believed the chances for life on the planet were "one hundred percent." Butler opted for the more conservative statement that "the planet is the right distance from the star to have water and the right mass to hold an atmosphere."

The announcement left the Swiss team collectively scratching their heads. They had previously found four small planets orbiting Gliese 581, including two borderline-habitable worlds hugging each end of the star's habitable zone. How had they missed two more planets? When they expanded and reanalyzed their own HARPS observations of the star, they readily confirmed once again the four planets their team had previously detected, but there was no signal of Zarmina's World or of the other potential planet, Gliese 581f. They argued that Vogt's less-precise HIRES data had introduced phantom worlds into their higher-quality HARPS observations of Gliese 581. Multiple independent examinations of the public HARPS and HIRES data came to divergent conclusions, with some finding evidence for the new planets and others dismissing them, depending on a shifting host of assumptions. If a certain variety of statistical analysis was used to extract the RV signals, only four emerged, rather than six—the new planets were false alarms! In their dynamical simulations, Vogt and Butler changed the orbits of the planets to nearly circular rather than moderately elongated, and found the results more stable with six worlds than with four—the new planets were real! But, adding up the various ways that the orbits of the four confirmed planets could interact, it appeared they could partially obscure and dilute RV signals of the two disputed worlds.

With uncertainty clouding the case of Gliese 581g, the prize of finding the first undisputed terrestrial world in a habitable zone remained up for grabs. In July of 2012, Vogt, Butler, and another colleague released a rebuttal to their critics, a two-steps forward, one-step back reanalysis of the HARPS data that scarcely mentioned Gliese 581f and modified Zarmina's World into a smaller 2.2-Earth-mass planet in a shorter 32-day orbit, still within the habitable zone. The potential world, they wrote, had approximately a 4 percent chance of being illusory, though since its signal was so weak much more data would be required for an airtight confirmation. To the Swiss, a 4 percent chance was still too high to bequeath confirmation upon Zarmina's World. Even a 1 percent chance would be considered borderline for such a planet. Extraordinary claims, they argued, required extraordinary evidence—evidence, they tacitly suggested, that would only come from a spectrometer at least as precise as HARPS.

In the summer of 2011, one of Butler's occasional collaborators, a thirty-two-year-old Spanish astronomer named Guillem Anglada-Escudé, began developing his own alternative data-analysis software to derive RV signals from HARPS spectral measurements, which the Swiss made public after a two-year proprietary period in accordance with ESO policies. Unlike the analytic methods of the Swiss, which threw away a significant amount of a star's raw spectrum, Anglada's software harvested a greater portion of a star's spectral data, extracting more signal from the noise to further boost RV precision for certain varieties of stars—particularly red dwarfs. Soon he was running his code on samples of the HARPS data, hoping to find borderline planetary signals that the less-precise Swiss analyses might have missed. His post-doc at Carnegie was almost finished, and he was looking for another job in the field—he thought a few planets under his belt could only help his chances. The first dozen data sets he examined came up empty of any new signals.

Late one August evening, after interviewing for a postdoc position at the University of Göttingen in Germany, Anglada returned to his hotel room and looked at another batch of HARPS data, 143 RV measurements taken between 2004 and 2008 for the red dwarf star GJ 667C, part of a triple star system some 22 light-years away from Earth. He fed his reduction of the data into Laughlin's Systemic software, and waited as the program looked for patterns. It first found the signature of a planet in a 7-day orbit that the Swiss had announced in 2009, but Anglada could see what looked like residual structure in the pointillistic clusters of measurements. He ran the data through Systemic again, and the software found a strong 91-day trend in the data—another possible planet, but also perhaps a cyclic stellar pulsation related to the star's estimated rotational period of 105 days. Anglada ran Systemic once more, nulling out the 7- and 91-day signals, then with trembling hands lit a cigarette and stared in disbelief at his laptop's glowing screen. Another sine wave snaked through the measurements, looking for all the world like the signal of a 4.5-Earth-mass planet, likely terrestrial, in a 28-day orbit firmly within GJ 667C's habitable zone. If the planet proved to be real, it would be assigned the name GJ 667Cc.

"It was very strange to find an unpublished, unclaimed, potentially habitable planet in a three-year-old public dataset," Anglada recalled to me. "So I looked again at the measurements using [the Swiss] method—the 28-day signal was there, but with a false-alarm probability that looked to be substantially greater than one percent." Too high, that is, to cross the HARPS team's traditionally ultra-stringent thresholds for announcing a discovery. Anglada's analysis of the 28-day signal, by contrast, yielded a false-alarm probability of three hundredths of 1 percent. He shared his findings with Butler, who excitedly agreed to gather more data on the star. Butler obtained twenty new RV measurements for GJ 667C, and Vogt supplied twenty older measurements from Keck's HIRES archives, both of which strengthened the 28-day signal. The team had soon modeled the putative system to examine its dynamical stability, and began drawing up a paper to announce the discovery. In the meantime, Anglada had decided he should bolster his case by seeking new data from HARPS, at the time still the best source in the world. Against the warnings of Butler and Vogt, who did not trust the HARPS team and urged him to publish with the data in hand, on September 28, 2011, Anglada submitted a proposal to ESO for twenty nights of HARPS observing time. The proposal did not explicitly announce Anglada's potential discovery but included GJ 667C on its short target list, as well as a figure discussing the star's 7-day, 91-day, and 28-day signals.

After submitting, Anglada closely watched the incoming traffic to his personal website, reasoning that he could gauge when his proposal was examined by looking for members of the ESO review committee who would visit to examine his credentials. In mid-November, his site received a pulse of traffic associated with computers in Munich, where the ESO review committee was based, as well as from computers in Geneva, Porto, Paris, and Santiago—all cities hosting HARPS team members. Each visitor browsed for minutes, then departed and did not return. On November 21, the HARPS team uploaded a 77-page preprint to a public online repository. The preprint was a draft version of a paper submitted to a prominent peer-reviewed journal, summarizing six years of HARPS observations, from 2003 to 2009. The astronomer Xavier Bonfils, a senior HARPS team member, was the first author. In a table on page three and a paragraph on page eight, the team noted its detection of a super-Earth in a 28-day orbit around GJ 667C, and referred interested readers to a more detailed forthcoming paper that was in the midst of preparation. Later, Anglada would learn his HARPS proposal had been rejected.

Vogt was the first to see the Swiss preprint claiming discovery of GJ 667Cc. He immediately sent a terse e-mail to Anglada and Butler:

"We've been scooped." Anglada was crestfallen; he read the preprint, then left his Carnegie office for a long walk. That night he couldn't sleep.

"I was very upset," he recalled. "So I reread the preprint again and started cataloging strange things. It didn't include any detailed analysis of the system's dynamics, and it barely discussed the 91-day signal. It said that the evidence for GJ 667Cc would be presented in a paper that was in preparation that would be submitted in future, yet it also said this paper had already announced GJ 667Cc—it was a self-contradictory statement that didn't seem to be a proper way to formalize the discovery." Looking at the table listing the planet's orbital parameters, Anglada noticed something odd. The table listed GJ 667Cc's orbital period as 28 days, but the size of its listed orbit erroneously corresponded to that of an orbit of 91 days, as if GJ 667Cc's entry had at one time concerned the 91-day rather than the 28-day signal.

"It could all have been coincidence," Anglada told me. "But I couldn't help feeling suspicious. If they had seen this signal in their data back in 2008, why did they wait three years, only to announce the planet in such a curious way the week after my HARPS proposal was reviewed? Why did the orbital size and period not match? I started to feel angry, and decided I should go on and push ahead with what I had found." Within a week, Anglada had completed his paper and submitted it to the *Astrophysical Journal Letters*, which published it in February 2012—beating the HARPS team to peer-reviewed publication. UC Santa Cruz issued a press release crediting Anglada, Butler, Vogt, and their collaborators with GJ 667Cc's discovery.

Bonfils and the rest of the HARPS team were aghast. They were the planet's true discoverers, they argued, as established by their November preprint. The controversy remained unaddressed until June, when Anglada and Bonfils agreed to a private meeting at a coffee shop outside a conference in Barcelona. Bonfils told Anglada that the HARPS team had already known of GJ 667Cc back in 2009 when they had announced their discovery of the system's other world in a 7-day orbit. They had submitted their 77-page survey paper for peer-reviewed publication in April of 2009, Bonfils said, but feedback from one of the reviewers had delayed the preprint's public release until November 2011. Anglada replied that the timing of the preprint was irrelevant, because it did not contain enough detailed information to support the HARPS team's discovery claims. Anyone could report wobbling stars, but to prove the wobbles were planets they had to show their analytical work. If published analysis was the test, Bonfils countered, then it was one Anglada's paper had still failed, because it contained the same mistake Butler and Vogt had made with the Gliese 581 system. Pooling HARPS data with that from lesser spectrometers such as HIRES, Bonfils maintained, would degrade the RV data and only increase the likelihood of false alarms; in contrast, the HARPS team's preprint was a valid discovery paper. By the time they finished their coffees, neither man had given any ground, and the tension between them had only grown.

I reached Bonfils by telephone a month after his meeting with Anglada. He sounded pained.

"They are trying to take credit for a discovery they did not make. It is as basic as that. It's not by hazard that we found this planet—it was on purpose. GJ 667C is one of our survey's most-sampled stars. That's why [Anglada] looked at it. HARPS was built by our team, and the scientific program and observations were done by our team. Most of the data reduction was already done and provided in our public data. I think it would be a pity if the guys who made the instruments and designed and performed the program of observations did not receive credit for their work. I'm a supporter of public data, but I had long feared someone would try to publish our data before us, and it has now happened. Right now this community rests on good behavior and gentlemen's agreements."

Nothing, nothing at all, Bonfils insisted, had precipitated the preprint's release in November 2011 after years of delay. It had been simple serendipity. "It was slow," he acknowledged. "I'm not proud of how long it took." He offered his own view of what had motivated the fight over public HARPS data. "Before, it was Marcy, Butler, Vogt, Fischer. Now they have split, the group has almost evaporated. I don't know Vogt or Butler personally, I have met Anglada once only. But I think there is, how to say, tension? You see it in their papers, in the language and accusations. A hunger, an aggression. I think it has become more difficult for them to get the observing time they need."

For the time being, Anglada told me, he had no plans to stop examining the public HARPS data for new discoveries. "People seem to think GJ 667Cc was a one-off thing, or that I got lucky looking where I did, but that's not true," he said. "This is really just the beginning of a bigger story. When you improve precision as I have, more things appear. The population of exoplanets is growing exponentially as we become sensitive to lower masses. I've looked at hundreds of systems now in their database. A lot of objects are showing up."

Despite working closely with Vogt and Butler, Laughlin had managed to stay above the battle for the first potentially habitable exoplanets. He had not been directly involved with the detections, announcements, and criticisms of Gliese 581g or GJ 667Cc, and he preferred to keep it that way. He held a longer view on the controversy surrounding them. In his mind, the strife between teams and the explosive expansion of exoplanetology were just growing pains, symptoms of a field struggling with its own imminent maturity.

"Just finding any planet around another star isn't as newsworthy or appealing as it used to be," Laughlin told me one afternoon in Santa Cruz. "That alone won't get you a flashy press conference and the front pages of newspapers and lurid artist's renditions like it would have ten, twenty years ago. Ten, twenty years from now, just finding an Earth-mass planet in the habitable zone of a Sun-like star probably won't be a big deal, either. Historians may look back and shake their heads at this period, when astronomers were regularly claiming to have found the 'first habitable planet,' but only in comparison to the last, previous 'first habitable planet.' It's my sense they'll remember this time as when the Heroic Age of extrasolar planet discovery came to a close."

"The real story," Marcy once remarked to me, "isn't the validity or the timing of discovery of any particular Earth-size, Earth-mass planet. Simply detecting one of these things does not overturn astrophysics or planetary science. The real story here is the amazing plausibility of detecting them at all, the fact that from our perch upon this speck of dust, we have come to the point where we are on the threshold of these sorts of discoveries. It's as surprising as an ant, living its life among other ants on an anthill, somehow calculating the size of the solar system. All we do is collect photons from the stars, and from that we can deduce the existence of planets and the scale and structure and future of the whole shooting match. It's crazy."

When, after repeated setbacks and delays, the APF at Lick Observatory finally became fully operational in 2013, observing time was evenly divided between the Marcy and Butler-Vogt teams. The break had been complete, and seemed irreversible: Butler and Marcy had not spoken since 2007, and perhaps never would again. And yet on nights when the sky above Mount Hamilton was dark and clear, they could be found virtually side by side, as their shared robotic telescope slewed between separate, distant points, building fractured empires among the stars.

CHAPTER 4

The Worth of a World

Back in 2009, less than a week after a Delta II rocket launched Kepler into planet-hunting history, Laughlin had quietly posted a strange, half-whimsical equation on his blog *systemic*, at oklo.org. In a series of subsequent posts, he explained how the long string of obscure variables and weighted functions could be used to crudely quantify the value of any terrestrial exoplanets that Kepler and the handful of other leading surveys might soon discover. It was, he said, an attempt to judge whether any particular "Earth-like" world was worthy of legitimate scientific excitement, independent of media hype. After plugging in a few key parameters—such as a planet's mass, its estimated temperature, and the age and type of its star—Laughlin's equation would generate a value, in

U.S. dollars, that could be assigned to that particular world. Small, rocky worlds in clement orbits around middle-aged, middle-of-the-road stars similar to the Sun merited the highest values, as those planets presumably offered the best chance for harboring complex biospheres that could eventually be detected by future space telescopes. For a planet to be worthy of wide attention, Laughlin opined, it would need to at least break the million-dollar mark.

Laughlin drew his economic baselines from simple math, dividing Kepler's federally funded $600 million price tag by 100, a conservative estimate of how many terrestrial planets the space telescope would discover during its lifetime. If such planets could be considered commodities, the math suggested that the 2009 market price, as determined by U.S. taxpayers, could be set at $6 million per world—a value that could drop over time if small rocky planets began to overflow astronomers' coffers. If, however, Kepler found a terrestrial world in the middle of a Sunlike star's habitable zone, Laughlin's test runs suggested such a planet's value could exceed $30 million in his equation. Zarmina's World, if it existed, garnered a value of around $60,000. GJ 667Cc was worth even less. According to Laughlin's calculations, Kepler's first million-dollar candidates appeared in February 2012. Several more would follow, bearing names such as Kepler-62f and Kepler-69c, until the Kepler spacecraft suffered a crippling malfunction in May of 2013 that all but ended its primary mission.

The cleverest part of Laughlin's valuation equation was its treatment of a planet's home star, which allowed his numerical scrutiny to be extended to the worlds in our own solar system. Photons, not dollars, are a planet hunter's fundamental currency, as they are what allow a planet to be not only detected but also subsequently characterized. Generally speaking, the more photons astronomers can gather from an exoplanetary system, the more they can learn about it. Stars and planets nearer to our solar system are brighter in our skies due to their close proximity, and hence more valuable, providing floods of useful photons where more distant objects would only offer trickles. This facet was why so many of Kepler's small planets would struggle to reach a valuation of even a million dollars: the Kepler field stars were far away, and thus very dim. The brightest star visible in the solar system by many orders of magnitude is, of course, the Sun, which has the capacity to send local planetary valuations into truly astronomical territory.

Based on the early-twentieth-century notion of Venus's clouds as reflective shielding against the potent solar flux, Laughlin's equation pegged the planet's value at one and a half quadrillion dollars—$1,500 trillion. Evaluating Venus based on its actual runaway-greenhouse surface temperature gave the planet a value of a trillionth of one cent. Laughlin sometimes compared such discrepancies in planets' values to the dot-com stock bubble of the mid- to late-1990s, when companies leveraged investors' irrational exuberance into billion-dollar valuations, only to crater when the bubble collapsed and their true, far lower values were revealed. When he ran his valuation equation for our own planet, Laughlin obtained a value of approximately five quadrillion dollars—roughly one hundred times the global gross domestic product, and, he reckoned, a handy approximation of the economic value of humanity's accumulated technological infrastructure. Searching for other habitable worlds, it seemed, was rather like speculating in a galactic-scale stock market.

Laughlin had also run his equation on a purely hypothetical Earth-size planet in the habitable zone of one of the two Sun-like stars in the Alpha Centauri system. He obtained a value of $6.5 billion—coincidentally about the same amount of money astronomers often estimate would be needed to build a space telescope capable of seeking signs of life on such a world. If humans actually voyaged there, Laughlin once pointed out to me, the star would become ever brighter, until it was a new Sun in a new sky of a New World. "So in going there, you have this ability to intrinsically increase value. And that's an exciting thing because it ultimately provides a profit motive for perhaps going out and making a go of it with these planets. This is saying that something that is several billion dollars on Earth could be, if you go there, a quadrillion-dollar payoff."

Months before our encounter at Tomales Bay, I had interviewed Laughlin about his equation for an article I published on the website BoingBoing.net. The article's contents made their way into the mainstream media, which focused far more on Laughlin's musing valuation of our world than on the worth of exoplanets. Stories appeared bearing headlines such as "Earth is worth £3,000 trillion, according to scientist's new planet valuing formula" (*Daily Mail*, February 28, 2011) and "Wanna buy the Earth? It'll cost you $5 quadrillion" (*Toronto Sun*, March 1, 2011). Angry e-mails began piling up in Laughlin's inbox, and television and radio stations called hoping to interview the mad scientist who so arrogantly placed a price on our planet. Laughlin was taken aback—he had emphasized, both in his posts and in his discussions with me, that his equation did not and could not assess the worth of, for instance, a human life or a new idea. Soon the story was churned out of view by the voracious 24/7 news cycle, but the sensational headlines left a lingering impression. Before Laughlin's talk at the Miller Institute symposium, I overheard one member of the audience jokingly refer to him as "The Man Who Sold the World."

The day after his presentation, I was in the front passenger seat of Laughlin's car as he drove us back down to Santa Cruz. In the back seat sat Taylor Ricketts, a World Wildlife Fund ecologist who had given a talk about "natural capital," the economic benefits of material goods and services provided by Earth's biosphere. Ricketts was part of a growing interdisciplinary push to study ecology in the context of economics, a field interested in not only the monetary value of, for instance, a pristine forest, but how that value might change if the forest was converted to pasture, or a parking lot.

At the time, Ricketts was a few months away from becoming director of the University of Vermont's Gund Institute for Ecological Economics, which he mentioned in passing not long after we crossed over the Golden Gate Bridge and began to drive on Highway 101 through downtown San Francisco. Gund's previous director, an ecologist named Robert Costanza, had "gotten into trouble" back in 1997 for a

Nature paper in which he tried to estimate the value of the planet, Ricketts said.

Laughlin's bushy eyebrows bounced up as he looked back at Ricketts in the rearview mirror. "What was the figure Costanza came up with?"

"Thirty-three trillion dollars per year, for all the world's ecosystems."

"I don't know why you'd get in trouble for that," Laughlin sighed.

"He made several basic economic mistakes that made his final figure essentially unsupportable," Ricketts said. "But more fundamentally, his critics just said, 'Thirty-three trillion dollars is a nice underestimate of infinity.' The value of the planet to us is infinite, because if all the ecosystems go away, life ends. For all of us. So there's not really a valid reason to put a number on that. Some people said he was silly for making his estimate, and others called him brave for trying. It's hard to know how much it has affected his career, but his name is kind of shackled to that paper."

Minutes passed. We eased into a snarl of afternoon traffic congested by a red stoplight at the crest of a long, high hill.

"So, an interesting counterargument to the 'infinite' value of the Earth is the fact that at some point this *will* all go away," Laughlin said. His eyes darted to sweep over the pedestrians slowly scaling the steep, tree-lined sidewalk, the cars idling their engines in the street, the people wandering in and out of boxy wooden row houses and tall office buildings of glass and steel, before his gaze finally came to rest back in the rearview. "And not because of anything we're doing, but because the Sun *will* evolve into a red giant and destroy the Earth. I don't think that's something for which we just have to sit down and acquiesce. So this becomes a question of where we are willing to begin talking about timescales on which our actions could have some conceivable utility."

"That's true," Ricketts said. "But economics is about how you make decisions under scarcity, right? You can't do everything, so how do you choose what to do and what not to do? You can't buy everything, so how do you choose what to buy and what not to buy? The reason to put value on things is to inform a choice you might have—that's the fundamental reason for economics to be. To place a value on the Earth . . ." He trailed off for a moment, finding his words. "I don't understand what the option or choice is there, what we would do with that information. It's not like we have the option of not being destroyed by the Sun, and that's probably why economists think a planetary valuation is a bit silly."

Laughlin shot a smiling glance my way. "But we do have an option. We can move the Earth."

A pregnant pause. "Move the Earth?"

"Sure."

"Like, tow it out of the way?"

"Essentially, yeah. We have more than enough time. Just take some large comets or asteroids from the Kuiper Belt and use them to tap and transfer some of Jupiter's orbital energy and angular momentum to the Earth over a timescale of hundreds of millions of years. Each time one flew by the Earth, you'd get a small kick, and you'd expand the Earth's orbit very gradually through those repeated close encounters. You'd need on order of a million close passes, one every several thousand years, but you could move the Earth's orbit out, close to where Mars is now. This is an idea I worked out ten years back with a couple of friends."

Another pause, as Ricketts pondered Laughlin's outlandish proposal. "That's cool. So it's a cost-benefit analysis: What would it cost us to tow the Earth out of the way, versus what's at stake on the benefit side?"

"You might destabilize and lose Earth's Moon," Laughlin said. "And you would have to be extremely careful to properly time each flyby so your object didn't collide with and sterilize the planet down to bacteria. But that intervention could net you billions more years for the biosphere, which is a *lot* of economic utility, and the cost is very small in comparison, because most of the energy to move the Earth is actually coming from Jupiter, transferred by the comet. You just have to

---

have very subtle control of the comet's trajectory when it's way out at the slow, far point of its orbit in the outer solar system. It's a matter of finesse more than brute force, but it's just rocket science. The point is, we could start doing this practically today, if we wanted."

"I'm still hung up on the value being infinite," said Ricketts, sounding skeptical. "It's all life on the planet. We know extinction is worth a lot of money to avoid without needing to quantify just how worthwhile it is."

"But there are hierarchies of infinity," I interjected. "Some are greater than others. The value of a commodity increases based on its scarcity, right? We still don't know how common Earth-like planets might be. Not just Earth-size balls of rock, but planets with water, weather, and living things. Maybe they're common out there, and life is cheap. But what if we end up taking the galactic census only to find that around the nearest five hundred or thousand stars there are simply—"

"No Earths," Laughlin finished, nodding. "We'll have to wait and see."

Debates over our planet's place in the universe and the cosmic worth of our world stretch back into time immemorial, to prehistoric speculations about the relationship between the heavens and the Earth that left their imprints on ancient myths and legends. By comparison, the oldest recorded scientific explanations for our cosmic context are quite young, though they still trace back some twenty-five centuries, to the towns and cities of Ionian Greece, scattered along the shores of the Aegean Sea.

In the sixth-century-B.C city of Miletus, in what is now southwestern Turkey, there lived an Ionian philosopher named Thales. Thales was the scion of a noble Phoenician family, and in his youth spent time in Egypt, where he learned geometry and studied ancient astronomical records. He was known throughout the ancient world for predicting a total solar eclipse that occurred over Central Anatolia on May 28,

585 B.C., but his greatest legacy is what we now call the "scientific method." Thales rejected the supernatural, teaching instead that rational thought and experimentation were the proper approach to making sense of the world. Thales believed everything in existence to be composed of one or more primeval substances and controlled by interacting forces—a belief that, in its essence, would be shared by any of today's particle physicists.

It was Thales's Miletian associate, Anaximander, who used these ideas to form a mechanistic explanation of the heavens. Anaximander believed that the universe was limitless and eternal. Beyond our realm, he said, far past where we could ever hope to see, other worlds endlessly formed and disintegrated within the boundless depths of an infinite void. Yet he also suggested that the Earth was at the center of our visible universe. Anaximander's Earth was a cylinder or disk fixed at a central point and surrounded by concentric shells that contained the fiery Sun, Moon, and stars. Pythagoras, Thales and Anaximander's young Ionian contemporary, thought instead that the Earth was a globe floating in space. He expanded Anaximander's model to include additional concentric shells for the planets, and held that the Sun, Moon, and planets orbited in perfect, harmonious circles about the Earth. If Thales and his disciples were the world's first true scientists, then Pythagoras and his followers were the world's first pure mathematicians: Pythagoras espoused the principle that numbers, whole and ideal, constituted the deepest reality, and that reality was best investigated not through the senses but through thought alone. The Pythagoreans' preference for mysticism and metaphysics was destined to prevail over Thales's empiricism to profoundly influence the philosophies of Plato and Aristotle some two centuries later.

Plato taught that the terrestrial world was made of four primal elements—earth, fire, air, and water—with a fifth element, the aether, forming the heavens. Aristotle incorporated some of these ideas into a larger cosmology, by stating that earth and water, being heavier than fire and air, fell and settled upon a central point, which became the

Earth. There could be no place like our own elsewhere in the heavens, because the heavens were made of an entirely different substance, and were perfect and unchanging. In the Platonic and Aristotelian reckonings, the Earth was privileged, corrupted, singular, and profoundly alone—a view that would dominate and stifle much scientific inquiry within Western civilization for nearly two thousand years.

It all could have turned out quite differently. In Plato's time, the staunchest defender of Thales's materialistic philosophy was an Ionian Greek named Democritus. For Democritus, the universe was built not from mystical numbers and geometric forms, but from infinitesimal physical particles moving eternally through the infinite void. Democritus called the particles *atoms*, the Greek word for "indivisible." Atoms and void, Democritus argued, were all that existed, and were thus the source of all things—including living beings and their thoughts and sensory perceptions. In a universe infinite in space and time, he said, the endless dance of atoms would inevitably lead to countless other worlds and other lives, all in an eternal process of growth and decay. Not all worlds would be like ours—some would be too inhospitable for life, and others would be even more bountiful than Earth. We should be universally cheerful, Democritus believed, at our fortune to exist in a welcoming world with so many pleasures. His constant mirth at humanity's tragicomic existence led his contemporaries to call him "the laughing philosopher."

Looking up at the dark Aegean sky, Democritus speculated that the stars, like everything else, were not made of a special celestial substance, but of atoms. They were simply suns, much farther away than our own, some so distant that in aggregate they formed the Milky Way's pale glow. Almost a century after Democritus's death, the idea of stars as distant suns reemerged in the work of another Greek astronomer, Aristarchus, who proposed that the Sun, rather than the Earth, was the center of our planetary system. By studying the size of the Earth's shadow upon the Moon during a lunar eclipse, Aristarchus surmised that the Sun was very much larger than our world, and felt it only natural that a smaller body

---

79

should orbit a larger one. Aristarchus further realized that his theory suggested the "fixed" stars were far more distant than most anyone had previously believed, based on a measurement called parallax. Parallax is the apparent motion of an object when viewed from a baseline defined by two separate points. You can see parallax very easily, by holding your finger in front of your face and gazing at it first with your left eye, then with your right, using your eyes as your baseline. Parallax tells you how far away an object is—the closer your finger to your face, the greater its parallax. Increasing the length of your baseline also increases an object's parallax—think of the apparent shift in the position of a lamp when viewed from opposite edges of one wall in a room. The fact that Aristarchus could discern no parallax shift among the stars when he viewed the night sky from opposite sides of Earth's orbit round the Sun told him that they were very far away indeed—light-years away, in fact, though that term had yet to be coined.

Aristarchus was accused of impiety for demoting Earth from its central place in the heavens; Plato reputedly so despised the ideas of Democritus that he wished for all the laughing philosopher's works to be burned. Ultimately, time alone would suppress what Plato could not, and for thousands of years the atoms of Democritus and the stars of Aristarchus would be largely forgotten. Only one minor work of Aristarchus has survived into the present day. None of Democritus's writings have passed down to us; we know of him through the writings of those he influenced, such as the later ancient Greek philosopher Epicurus. Epicurus's work, too, is mostly lost. A good portion of what we do know of his philosophy comes from a single book-length poem, *De Rerum Natura* (*On the Nature of Things*), written in Latin hexameter around 50 B.C. by the Roman scholar Lucretius. In the poem, Lucretius praised and summarized Epicurean thought—including the idea of atoms, an infinite universe, and the inevitability of other living worlds. It did not stand to reason, Lucretius wrote, that "this was the only world and heavens created, and that beyond it those many bodies of matter do nothing at all." Atoms, numberless in the totality of infinite space, would on occasion coalesce to "become the beginnings of great things: of the world, sea, sky, and the race of living creatures. . . . It must be admitted that there are other worlds in other regions, as well as different races of men and breeds of wild beasts. . . . In the sum total of things, there is nothing singular, which is born unique and grows unique and alone; it instead belongs to some class, and there are very many of the same kind."

Lucretius's poem would have perhaps been lost as well, if not for a single worm-eaten copy that was found by an Italian book collector in the dusty depths of a German monastery in 1417. Soon after the discovery, the poem was translated into modern languages, printed on Gutenberg presses, and distributed throughout Europe, where it and other ancient rediscovered works helped foment the scientific revival of the European Renaissance. It would be well more than a century, however, before the renewed idea of Earth's modest place in an infinitude of worlds would make its greatest impact. The revolution began with the 1543 publication of *De Revolutionibus Orbium Coelestium* (*On the Revolutions of Heavenly Orbs*) by the Polish cleric Nicolaus Copernicus, laying out a heliocentric model of the solar system. The work consumed the last thirty years of Copernicus's life, and he received the first printed copy while on his deathbed. Copernicus, like Aristarchus nearly two millennia before him, had shown that the apparent motions of the planets in the sky could be more elegantly explained if the planets were moving around the Sun rather than the Earth.

In 1610, the Italian astronomer Galileo Galilei confirmed the Copernican model after turning the newly invented telescope to the heavens. He glimpsed sunspots on our star, and from their gradual motion deduced that the Sun, like the Earth, was spinning. Jupiter, he found, was circled by a number of smaller moons, confirming that smaller bodies do indeed orbit larger ones, and that not everything orbits only the Earth. When he gazed at Venus over the course of a year, he saw the planet swing through a full set of phases, just like the waning and waxing Moon—evidence that Venus passed both in front of and behind the illuminating Sun. This was the crucial evidence for heliocentrism, as geocentric models predicted that Venus, closer to the Sun but still circling Earth, would consequently always be backlit by the Sun's radiance and would display only a crescent phase as viewed from our world. Yet the Copernican model was still imperfect: it failed to replicate the exact motions of the planets in the sky. In making his model, Copernicus had implicitly relied upon the old Pythagorean notion, extended by Plato, Aristotle, and others, that planets moved in perfect circles.

Around the same time that Galileo began using his telescope, a German astronomer, Johannes Kepler, announced a discovery that would mark the true beginning of modern astronomy, one that ironically came to him while he worked to create a table for casting more accurate horoscopes. Kepler had been grappling with the orbit of Mars, trying to integrate historical records of the planet's motions into Copernicus's heliocentric model. He had painstakingly considered circular orbits, even spiraling orbits, but his results did not align with observations. At last, on a despondent whim, he decided to treat Mars as if it moved in a squashed, elongated oval, an ellipse—an idea so basic that Kepler assumed it had already been investigated by prior generations of astronomers. To his great surprise, the results of his calculations beautifully mirrored observations. He subsequently confirmed that the orbits of the other known planets were elliptical as well, and he went on to use the revelation of elliptical orbits to codify his three laws of planetary motion.

The first law simply stated that each planet moves in an ellipse, with the Sun at one focus. The second law was that planetary orbits sweep out equal areas over equal times. This means that when a planet's orbit brings it closest to a star, the planet must move faster than it does at the opposite end of its orbit in order to sweep out the same area over the same period of time. The third law stated that the square of a planet's orbital period is directly proportional to the cube of its average orbital separation from the Sun, establishing a clear relationship between the length of a world's year and its distance from its star. This explained why Mercury, closest to the Sun, raced across Earth's skies, and why Jupiter and Saturn, much farther out, moved so sedately. Kepler's third law allowed him to estimate the proportional distances of the planets: he determined, for instance, that Mars was one and a half times the distance from the Sun that our Earth was, and that Jupiter was more than five times as far, though the actual distance from the Earth to the Sun remained unknown.

The importance of Kepler's findings cannot be overstated. Near the end of the 1600s, Isaac Newton would use Kepler's laws to derive the universal laws of gravity. Today, knowledge of Kepler's laws is what allows mission planners to chart the courses for interplanetary spacecraft, and is how planet hunters determine whether an exoplanet resides in its star's habitable zone based on the world's orbital period alone. Kepler in a sense unified the heavens and the Earth, showing without a doubt that they existed within one framework dictated by the same physical laws. He also gave solid theoretical heft to what became the crux of Copernican thought: that, all things being equal, the Earth, and by extension the solar system, could not be assumed to be singular, unusual, or in any way privileged, but rather should be presumed common, mediocre, and average—at least until evidence proved otherwise. This "Copernican Principle" or "principle of mediocrity" has quietly guided physics, cosmology, astronomy, and planetary science ever since, though not always in the right direction. Galileo, viewing the Moon's patchy, cratered terrain through his telescope, declared it to be a world like ours of lands and seas. Kepler even speculated that the Moon was inhabited, and thought some of its more intelligent creatures might have excavated the circular lunar craters to house their cities. Whether in dreams of jungled, primitive Venus or mirages of an advanced, dying civilization building canals on dried-up Mars, for centuries it was common for learned and reputable scientists to vocally profess that most and perhaps even all worlds were habitable and inhabited.

In 1627, while using his new laws to calculate the future motions of Venus, Kepler surmised that the planet would occasionally cross the face of the Sun as viewed from Earth. He calculated that the next transit of Venus would occur over the course of several hours on December 6, 1631, and that other than a near miss for a transit in late 1639, Venus would not cross the Sun's face again until some time in 1761. Kepler hoped to witness the 1631 transit, but died in 1630. The transit of 1631 came and went, apparently unseen. In 1639, scarcely a month before the near miss Kepler had calculated, the young British astronomer Jeremiah Horrocks discovered an error in Kepler's calculations—Venusian transits actually occurred in pairs separated by 8 years; the interval between each pair oscillated between 121 years and 105 years. Horrocks calculated that on the afternoon of December 4, 1639, the transit of Venus could be seen from his home in northern England. He and a friend, William Crabtree, rushed to plan their observations. On the fated day both men watched an event no human eyes had ever before seen, as the silhouette of Venus, one-thirtieth the apparent diameter of the Sun, glided across the blazing star. They were the only two souls on Earth to witness the 1639 transit. Horrocks's correction of Kepler's calculations set the timing for transits in future years. A pair would occur in 1761 and 1769, then in 1874 and 1882, then in the far-off years of 2004 and 2012, continuing on and on in what was thought to be an endless cycle.

Writing in the *Proceedings of the Royal Society* in 1716, the English astronomer Edmond Halley suggested how Venusian transits could provide an absolute Earthly reference point against which the rest of the universe could be measured. When viewed from different places on Earth, Halley wrote, the path of Venus across the Sun would shift slightly, also shifting the transit's duration. By precisely timing the transit to distinguish the shift between two widely separated locations, it would be possible to triangulate the distance between the Earth and the Sun. From there, simple math would yield the Sun's true size and each planet's orbital distance, revealing the physical breadth of the solar system. In the years leading up to the next transit, that of 1761, states across Europe organized more than a hundred teams to travel to the world's far corners to attempt Halley's proposed measurements. It was the first-ever flowering of international, state-sponsored science, and it was a spectacular failure. Astronomers hauled delicate equipment by ship, sled, and horseback into wild areas where the transit would be viewable, often only to find their cargo shattered and warped beyond repair at their destination. Wars, disease, and poor weather scuttled many attempts well before the transit actually occurred. The measurements that did trickle back from far-flung expeditions were too inaccurate and contradictory to be useful.

Of all the astronomers who sought to study 1761's Venusian transit, none were unluckier than Guillaume Le Gentil of France. Le Gentil left his home in Paris a year before the transit, bound for a French colony in India. After he left, war broke out between France and Britain, and his ship was blown far off course by a storm. When he finally arrived in Indian waters a few days before the transit, he was barred from coming ashore by British troops, who had seized the French colony. Le Gentil was forced to observe the 1761 transit offshore, where heaving seas made precise measurements impossible. He stoically remained in Asia to await the next transit. After eight patient, painstaking years, by 1769 Le Gentil had constructed a small observatory in India to record the event. All was ready and the weather was fair on the eve of the appointed day, June 4, 1769. A thin haze ominously accumulated overnight, then boiled off in the morning sun. Moments before Venus was to begin its passage across the Sun, a thick bank of clouds rolled in. They dissipated only later that afternoon, shortly after the transit's conclusion. Le Gentil was briefly reduced to a gibbering mass of twitching nerves, but after a time regained his senses and began his long journey home. His homeward voyage was derailed first by dysentery, then by a hurricane that nearly sunk his ship. Arriving empty-handed back in Paris in 1771, eleven and a half years after he had left, Le Gentil found his former life in tatters: he had been declared legally dead and his estate had been dissolved.

Some were more fortunate than Le Gentil in 1769. From a hilltop in Tahiti, Captain James Cook successfully charted the transit of Venus for the Royal Navy and the Royal Society before going on to map and claim islands for the Crown on his voyages throughout the South Pacific. On his farm in Philadelphia, an astronomer named David Rittenhouse documented the transit for the American Philosophical Society, bringing the burgeoning colonial scientific community onto the world stage for the first time. As an astronomer, Rittenhouse was arguably already of somewhat delicate temperament, and was so overcome by the transit's first moments that he fainted, leaving an otherwise inexplicable lacuna in his official records. Combining these and other measurements from expeditions scattered across the globe, astronomers pegged the Earth-Sun distance, the Astronomical Unit, at 150 million kilometers, or 93 million miles. At last, astronomers had a firm foundation for calibrating the size of the solar system, and with it, the universe. The Copernican Revolution could proceed.

Now knowing that the Earth drew out an approximately 186-million-mile baseline as it moved in its orbit around the Sun, astronomers revisited the ancient parallax measurements of Aristarchus and began measuring the distances to the stars. Over months and years, a handful of nearby stars revealed their proximity by moving against the more distant "fixed" stars, just as a low-flying bird would whiz through your field of view against the more stately motion of a passenger jet flying overhead much farther away in the sky. By the middle of the nineteenth century, astronomers were regularly measuring stellar parallaxes, establishing that most stars in the sky were, at minimum, tens of light-years away. Our own solar system seemed caught in a cycle of perpetual demotion, occupying an ever-shrinking region within a universe that grew with each new improvement in measurement.

In the first decades of the twentieth century, American astronomers built off the basis of stellar parallax to enact the next great Copernican demotions, establishing the field of modern cosmology. First, the spatial distribution of the Milky Way's star clusters revealed that our solar system was not, as many had believed, at the center of the galaxy, but rather on its outskirts. Then, the American astronomer Edwin

Hubble found that our galaxy was but one of many, and discovered that nearly all other galaxies in the sky were racing away from one another at incredible speeds. The universe was literally expanding, following a course that would soon be elucidated in the relativistic theories of Albert Einstein. Once again, at the largest scales that could then be measured, the cosmos was proving far larger and stranger than most anyone had dared to previously suppose, with our existence nowhere near the center.

Meanwhile, far back down the scale, in the realm of stars and their planets, the Copernican Revolution had stalled. Astronomers mapping nearby stars had gradually discovered that our Sun was not a typical star at all—most of its neighboring stars were smaller and dimmer, red and orange dwarfs. Perhaps the solar system was atypical as well, since no solid evidence for exoplanets had been obtained. Many astronomers began to believe our Sun might harbor one of only a very small number of planetary systems in the entire galaxy, though by the middle of the twentieth century mounting indirect evidence suggested planets were probably common around stars.

Still, the Space Age's chilling revelations about Venus, Mars, and the solar system's other apparently lifeless planets gave Earth a small fraction of its previous Platonic luster. Then came the exoplanet boom. To many modern planet hunters, finding another biosphere beyond the solar system became a quest for a comforting capstone to place atop the principle of mediocrity, forming the pinnacle of the Copernican Revolution. At last, our planet and all upon it would reach its final demotion— just another average world in a cosmos teeming with life.

And yet, leaving aside the vexing unsolved mysteries of life's origins and the unknown quantity of Earth-like planets, the frontiers of cosmology have recently unearthed new difficulties with the Copernican Principle's notions of our mediocrity. The majority of the observable universe looks to be empty space, offering at best one-in-a-million odds that, set down randomly within it, you would find yourself in a galaxy. Given that the universe is gradually expanding, these odds can only get worse as time marches on. Mysterious halos, filaments, and clouds of "dark matter," seemingly immune to all forces in the universe save for gravity, are what hold galaxies and galactic clusters together. A galaxy's interior is mostly void, filled with, on average, one proton per cubic centimeter. If a galaxy's stars were the size of sand grains, the average distance between them would be on the order of a few miles. Only the slimmest fraction of the interstellar material within a galaxy is at any moment condensed into something so sophisticated and advanced as a hydrogen atom. To simply be *any* piece of ordinary matter—a molecule, a wisp of gas, a rock, a star, a planet, or a person—appears to be an impressive and statistically unlikely accomplishment.

The apparently privileged place of matter within such vast emptiness is compounded by the universe's ongoing evolution, which seems set on a course toward ever-greater desolation. Surveys of supernovae detonating at the fringes of the observable universe have revealed that the space between galaxies is not only expanding, but also accelerating in its expansion, propelled by a mysterious force cosmologists know only as "dark energy." Unless somehow the cosmos ceases its accelerating expansion, the universe of the very distant future will be far lonelier and emptier than it is now: other than a handful of galaxies gravitationally interacting with the Milky Way, known as the "Local Group," all the other galaxies we presently see in our skies will by that late date have been swept beyond the horizons of our visible universe. The Local Group's galaxies will also eventually become dark some hundred trillion years from now, as all their stars burn out one by one. Next, decillions upon decillions of years in the future, protons—the cornerstones of atomic structure—should all decay in dying bursts of radiation (a "decillion" is one followed by thirty-three zeroes, and a very, very long time indeed). As this process occurs, the last remnants of burnt-out stars and frozen planets will dissolve into oblivion. The universe will become incomprehensibly dark, diffuse, and cold, and in the minuscule sector that used to harbor our Local Group, the only remaining macroscale structures will be a few supermassive black holes, slowly evaporating due to quantum-mechanical effects. When at last the last black holes shrink and vanish in puffs of quantum foam, there will be little left but faint wafts of photons, electrons, and neutrinos streaming endlessly through the infinitely expanding emptiness.

Perhaps it is simply a failure of imagination to see no hope for life in such a bleak, dismal future. Or, maybe, the predicted evolution of our universe is a portent against Copernican mediocrity, a sign that this bright age of bountiful galaxies, shining stars, and living planets, unfolding only a cosmic moment after the dawn of all things, is in fact rather special.

Just as the universe's future challenges Copernican expectations, so too does its past. The essential idea behind the Big Bang, the leading scientific explanation for the universe's past history, is that the cosmos developed from a singular, improbably dense point that somehow explosively expanded about 13.8 billion years ago. Not very Copernican. More problematically, the Big Bang itself is challenged by the universe's structure. Beyond the granular distinctions of atoms, planets, stars, galaxies, and galactic clusters, on the largest scales astronomers can measure the cosmos appears preternaturally smooth. This large-scale smoothness is in keeping with Copernican predictions, but vexing, since even the slightest difference in expansion rates between separate regions of the early universe should have resulted in substantial deviations in their present structure—lumps, wrinkles, and the like. But regions of space now at opposite sides of the observable universe appear structurally identical, almost flawlessly smooth despite being so distant from each other that they are causally disconnected. Light itself has yet to travel between them, not to mention any information or energy or heat that could bring those far-removed sectors of the universe into equilibrium.

The leading cosmological explanation for this conundrum is an add-on to the Big Bang called "inflation," which posits that fractions of a second after our universe's birth, when everything was squeezed into a hot, dense region perhaps the size of a proton, an intense, mysterious blast of repulsive anti-gravity suddenly "inflated" the space to perhaps the size of a large grapefruit. This may sound minor, but it represents a leap in scale of some ten trillion trillion. Any major irregularities would have been erased by this accelerated expansion, like the creases that disappear from the rubber surface of an inflating balloon. According to inflationary models, the minor imperfections that remained came from vastly magnified quantum fluctuations, and formed slight pockets of density in the early universe from which galaxies and galactic clusters condensed.

The problem with inflation is that once it begins, it cannot be easily stopped. Some researchers have even speculated that dark energy may be a bizarre echo or shadow of primordial inflation, somehow returning after billions of years of dormancy. Though primordial inflation may rapidly decay and cease in a local region of space (such as our entire observable universe), because it so greatly boosts the rate of expansion, it should thrust a vastly larger bubble of space out far beyond the horizon of our visible universe. Indeed, a universe expanded far beyond our observable universe's horizon is a standard outcome of primordial inflation. Deep within that exponentially larger, perhaps infinite volume, more inflationary Big Bangs could then occur again and again even if they were extremely improbable. Each time, yet another branching expansion without end would emerge. Inflation, once started, seems set to proceed eternally, generating an infinite, fractal ensemble of parallel bubble universes, each related to but causally distinct from the others. Most would be destined never to intersect and meet, as ongoing inflation in the spaces between them moved them apart faster than their boundaries expanded, like fleeting bubbles in a white-rushing river. Within different bubbles, the laws of physics that froze out from the fiery chaos of inflationary expansion could be utterly different than those that reign within our own local region of the universe.

In some small fraction, the laws of physics would be identical to or scarcely distinguishable from our own, and those regions would be more likely to generate galaxies, stars, planets, and living creatures. In the remainder of the regions, natural laws would be so alien that life as we know it would be impossible. The theory of an inflationary "multiverse" is consequently often used in modern cosmology to explain otherwise mysterious fundamental properties of our universe that seem fine-tuned to allow life to arise and persist. In some stillborn universes with physical laws that precluded life's emergence, there would be no stars. In others there would be no atoms. Some would expand or contract so quickly they shuffled out of existence in an instant; others would contain exactly equal portions of matter and antimatter that would mutually annihilate in a blaze of energy, leaving behind nothing but vacuum and seething radiation fields. In the majority of universes we can conceive of, the existence of observers seems inconceivable. There would be no living creatures within them to gaze out at their surroundings and wonder how it all began. In this telling, the universe we see around us is of course fit for life, for otherwise we would not be here.

No one has yet devised a foolproof way to test most of these ideas—how, exactly, can you detect other universes that by definition are forever inaccessible to us? But if true, an inflationary multiverse holds muddled consequences for Copernican ideas. On the one hand, it would mean that our entire observable universe was only the most minuscule fraction of a much larger cosmos inflated from our Big Bang 13.8 billion years ago. This vastly larger cosmos would itself be only a single member of an infinite ensemble of other universes. Infinity being, well, infinite, it would follow that the multiverse would host infinitudes of living beings on a limitless number of other worlds. On the other hand, the infinitude of bubble universes incapable of supporting life would appear to be very much larger than the infinitude that could. Against the principle of mediocrity, an inflationary multiverse suggests our local universe is a small part of an atypical bubble embedded in a much larger region of inflation, a member of a rather exclusive subset of universes that can harbor life. Whether the physical laws we observe are "average" within this subset, no one can say. A planet, a star, or a galaxy may be only as special and valuable as the cosmos that gave birth to it.

In contemplating eternal inflation, modern cosmology has, in effect, returned to some of the tenets first formulated by the Greek atomists some 2,500 years ago; Democritus would certainly laugh that it took so long. In the far future, as our own universe decays into a dark, cold senescence, life's last holdouts may find some solace in believing that somewhere, far, far away, over the cosmic horizon, the ceaseless process of creation continues, giving birth to new lives, new worlds, and new universes. Hope springs eternal.

CHAPTER 5

After the Gold Rush

From time to time when Laughlin was deep in thought at his office, he would absentmindedly reach across his desk for a small child's toy he had purchased in the 1990s, back when he was a postdoc at UC Berkeley. The toy looked much like a hangman's scaffold. Instead of a noose, the scaffold held a thin steel pendulum, loosely suspended above a steel square by a tiny embedded magnet. He would place magnets of various strengths and shapes strategically upon the square and give the pendulum a gentle bump; it would swing to and fro for long periods, kicking between magnetic fields with sufficient force to overcome the frictional loss of momentum from moving through the air. Its motions followed a chaotic random walk, never exactly repeating any given path.

Laughlin savored the toy for how its complex behavior could unfold solely from the simple initial conditions of each magnet's position and the strength and trajectory of an initiatory nudge. It reminded him of his struggles to predict the typical outcomes that emerged from the chaotic gravitational interactions of black holes, stars, and planets, and his efforts to squeeze faint signals from backgrounds of meaningless noise.

One night near the end of June in 2006, after coming home late from work, Laughlin sat down at his kitchen table and realized he had brought his work home with him—an idea was effervescing in his brain. Earlier that day he had been pondering the uncertain orbit of Proxima Centauri, a dim red dwarf tenuously bound by gravity to Alpha Centauri's binary system of two Sun-like stars, Alpha Centauri A and B. Whether it was only a lone star passing in the galactic night or an estranged family member of the Alpha Centauri system, Laughlin was not sure. What mattered was that the trio comprised the closest known stars to our solar system. As he had thought that day about the stars' celestial motions, the question of whether they had any accompanying planets intermittently tickled the back of his mind. By that night, the tickle had become an irresistible itch. Laughlin scratched it with notes scrawled on scrap paper and calculations keyed into his laptop.

For decades, consensus had held that binary star systems were poor targets for planet searches, because it was thought that gravitational interactions between the two stars would either prevent planet formation or fling planets, once formed, on escape trajectories out of the system. But ever since the exoplanet boom, increasing numbers of binary-star planets had been discovered—the consensus had been wrong. Due to their close proximity to Earth, Alpha Centauri A and B offered plentiful photons for an RV planet search. Alpha Centauri B, a dusky orange star slightly smaller than our Sun, was particularly quiet and stable—an excellent candidate to scour for potentially habitable planets. Earlier searches had already ruled out the presence of any gas-giant planets within a few AUs of each star, but the presence of smaller worlds was still possible. Laughlin thought they might be just within reach.

No matter how many holes Laughlin tried to poke in his thinking, deeper scrutiny deflected each of his criticisms, and the idea of a Centauri-centric RV survey stood unscathed. The more he turned it over in his head, the more ideal and fortuitous the situation seemed. Though most stars appear immobile in the sky on human timescales, the Sun's 250-million-year orbit about the Milky Way's center ensures that every few hundred thousand years our solar system has entirely new neighbors. "If we were plopped down at some random point in the galaxy, there's only a 1 percent chance we'd find ourselves near stars so optimal for detecting small rocky planets like our own," Laughlin told me during an interview in late 2008. "The hand of fate has dealt us a very interesting situation that has not existed for at least 99.9 percent of Earth's history. It's remarkable that Alpha Centauri is right next door just as humans emerge and develop the ability to make these measurements. I'm enamored with that coincidence."

A survey seemed worth the risk of coming up empty-handed, Laughlin told himself as he sat at his kitchen table on that summer night in 2006. Finding any planets at all around the stars of Alpha Centauri would be a historic discovery. By virtue of their close proximity they would be prime targets for subsequent study, and regardless of their characteristics would likely garner large amounts of in funding for further research.

For a moment he allowed his thoughts to drift far away, into hazy realms of possibility. Detecting potentially habitable worlds around Earth's nearest neighboring stars would be a truly revolutionary development, one that could stimulate major investments and advances in the quest to learn about our place in the universe. Without actually *seeing* any terrestrial planets that orbited in the stars' habitable zones, no one could know whether they were places like Venus, Mars, Earth, or something else entirely outside of expectations. In comparison with the prospect of confirming other living worlds right on our galactic doorstep, the cost of building a direct-imaging space telescope to study such planets would seem to shrink. If, by happy chance, those nearby worlds looked particularly inviting when viewed through new tele-scopes, they would lure generations of scientists, explorers, and dream-ers just as the planets of our own solar system did during astronomy's earlier, more romantic eras. Alpha Centauri would call across that briefest interstellar gulf, and someone would surely strive to answer. The first emissary would undoubtedly be robotic, maybe something the size of a Coca-Cola can somehow sent voyaging at 10 percent of the speed of light. And if, nearly a half century after its launch, the probe against all odds beamed back to Earth high-resolution images of another clement planet replete with oceans, clouds, continents, and . . .

Laughlin blinked and reined in his mind's free roaming, which had suddenly slipped beyond the stars. Too much extrapolation was dangerous. He closed his laptop, rose from his kitchen table, and went to bed.

Within months of his kitchen-table reverie, Laughlin had per-formed numerical simulations of planetary assembly in Alpha Centauri with a graduate student, Javiera Guedes. They began with Moon-size planetary "embryos," and watched as the embryos gravitationally clus-tered into small, rocky planets in stable, habitable orbits around each star. Laughlin next approached Debra Fischer, Marcy and Butler's for-mer collaborator, to propose a search. With funding from the NSF and help from many colleagues, including Laughlin, Butler, and her own students, Fischer began an intensive survey of Alpha Centauri in 2009, using a small 1.5-meter telescope at the Cerro Tololo Inter-American Observatory in Chile. Sixty kilometers to the north, at Cerro Paranal, the Swiss had been monitoring Alpha Centauri B since 2003, but soon after Fischer's program began, they drastically upped the cadence of their observations. They could not focus as intently on the stars as Fischer and her collaborators—HARPS was simply too valuable a re-source to be monopolized by a single star system. In 2011, a third team officially joined the search, obtaining funding to perform a high-cadence survey using a 1-meter telescope at Mount John University Ob-servatory in New Zealand.

Analyzing the RV data proved harder than anticipated, partially due to difficulties with precisely removing the binary orbits—Alpha Centauri A and B orbit each other in a roughly 80-year period, at an average separation somewhat greater than the distance between our Sun and Uranus. The orbit is significantly "eccentric" (non-circular), but its fine details were not known down to the level of centimeters-per-second, making RV signals of small planets in either star's habitable zone harder to see. That eccentric orbit posed further problems, according to another round of numerical simulations performed by the theoretician Philippe Thébault of Paris Observatory and a few collaborators. In run after run, gravitational perturbations driven by the orbit's eccentricity disrupted the formation of structure well before any Moon-size building blocks could coalesce. Thébault's simulations suggested nothing larger than sand grains and pebbles would orbit either star.

Laughlin could find no potential oversights or errors in the newer simulations, save for one: Thébault had assumed that the stars were born at their current distances from each other, with each sporting a protoplanetary disk approximately the same size as the one astronomers think formed our own solar system's planets long ago. Laughlin believed Alpha Centauri's stars had begun with very different initial conditions, at wider separations, perhaps with stubbier, smaller protoplanetary disks—any of which could prevent Thébault's subsequent eccentric disruptions. The presence of the red dwarf Proxima was a potential piece of forensic evidence, Laughlin thought. "Had Alpha Centauri formed in a very dense stellar cluster, like what we see in the Orion Nebula today, then in all likelihood Proxima would have been stripped from its orbit by a passing star," he explained to me. "Of course Proxima may have been only captured much later on, but I would bet its presence means Alpha Centauri formed in a more open, less cluttered environment, where the stars were further apart. . . . If you start with [Thébault's] initial conditions, you'll end with no planets. I just believe the initial conditions were quite different than what he uses."

In October of 2012, a discovery was finally made. Using more than 450 combined HARPS measurements, the Swiss had detected what looked to be an Earth-mass planet around Alpha Centauri B. It resided in an inhospitable three-day orbit, so close to the star that its surface would be broiled at temperatures exceeding 650 degrees Celsius, yet it was universally acclaimed as a promise of great things to come, of a life-friendly cosmos. Somehow, even in uninviting circumstances, small, rocky planets still found ways to form and persist. Alpha Centauri Bb, as the new world was called, is so light it only creates a wobble of some 50 centimeters per second upon its star—a bit faster than the average speed of a crawling baby. If HARPS could detect that faint signal, the Swiss said, then any undiscovered rocky planets in Alpha Centauri B's habitable zone were also likely within reach. And more planets were almost certainly there—statistics from the Kepler mission strongly suggested that where one small, close-in planet was detected, several more would lurk farther out, as yet unseen. Astronomers began to murmur that, in all likelihood, all three of Alpha Centauri's stars possessed planets, and that quiescent, placid B would again be the first to yield additional discoveries. It was only a matter of time.

Once, years before its first planet was detected, I asked Laughlin via e-mail if he had ever dreamed about the environment of Alpha Centauri. Did he ever try to visualize what a habitable world might look like around one of Alpha Centauri's stars? His reply contained only a passage from *The Martian Chronicles* by Ray Bradbury:

The old Martian names were names of water and air and hills. They were the names of snows that emptied south in stone canals to fill the empty seas. And the names of sealed and buried sorcerers and towers and obelisks. And the rockets struck at the names like hammers, breaking away the marble into shale, shattering the crockery milestones that named the old towns, in the rubble of which great pylons were plunged with new names: IRON TOWN, STEEL TOWN, ALUMINUM CITY, ELECTRIC

Village, Corn Town, Grain Villa, Detroit ii, all the mechanical names and the metal names from Earth.

Laughlin declined to elaborate, but the quotation suggested we would unavoidably filter any alien world's mysteries through our familiarities, reshaping all we found into our own image, our initial conditions from Earth.

Months later, out of curiosity I sought Laughlin's opinions on SETI. Would it ever be successful? He laughed coolly, then began thinking aloud. "Heh. Maybe. Ahh . . . Well. Eventually. Though not how most people think. If you get a radio signal, that's great, you can get right down to work. That's a nice dream. That works well for some fraction of our galaxy. A big space telescope could look for some signs of life on planets in the habitable zones of some very nearby stars, but distinguishing signs of intelligence would be tough. I think if a SETI detection ever comes, it will probably be extragalactic."

Somewhere out there, Laughlin said, perhaps beyond our present observable horizon, perhaps in another universe entirely, there would be galaxies containing civilizations very much like ours, except with much more fortunate initial conditions. Maybe they had emerged, as we had, on a planet like Earth, with a moon close and tempting in the sky. Maybe their star had not one, but two or even three habitable Earth-like planets, and their nearest neighboring star had held habitable worlds as well. Maybe they had originated in a binary or trinary star system, with each sun blessed by multiple habitable worlds. If a civilization somewhat like ours got lucky enough, he said, and took advantage of its good fortune with the kind of "mid-twentieth-century-style Space Age expansion" that had flashed and faded here in our own solar system, it would reasonably have a chance of spreading out to grasp and shape its entire galaxy. Evidence for such galactic empires could be lurking latent in any number of the thousands of unnamed galaxies contained in a typical long-exposure "deep field" image from current space telescopes.

I wondered what that evidence might look like. Laughlin laughed again and said that was something he couldn't predict.

On July 13, 1963, just off the Cabrillo Freeway in San Diego, a time capsule was sealed in a small subterranean concrete vault beneath what was then the west entry ramp into the General Dynamics Astronautics plant, where the company built Atlas rockets for the U.S. government. General Dynamics was bought out in the 1990s, and much of its Atlas-manufacturing infrastructure was dismantled to make room for easier money from industrial parks and office buildings. The capsule, meant to be unearthed a hundred years after its burial, was instead dug up and relocated to storage at the San Diego Air & Space Museum in Balboa Park. If you were to open the capsule today, you would find a slim, aged volume entitled *2063 A.D.* The book had been commissioned to commemorate General Dynamics's fifth anniversary, and contained hopeful prophecies from experts—generals, politicians, scientists, and astronauts—about humanity's conquest of space a century hence. Someone at the company thought to print a few hundred extra copies, which is how we know about the book's contents today.

Mercury astronaut John Glenn, the first American to orbit the planet, predicted that within a century we would have linked atomic power plants to "anti-gravity devices," fundamentally rewriting the laws of physics and revolutionizing life and transportation on Earth and in the heavens alike. Another Mercury astronaut, Scott Carpenter, expressed his hope that the anti-gravity "scheme" would help humans colonize the Moon, the Martian moon Phobos, and Mars. The prominent astronomer Fred Whipple suggested that Earth's population would have stabilized at 100 billion, and that planetary-scale engineering of Mars would have altered the Red Planet's climate to allow its 700,000 inhabitants to be self-sufficient. The director of NASA's Office of Manned Space Flight, Dyer Brainerd Holmes, suggested that in 2063 crewed vehicles would be reaching "velocities approaching the speed of light," and that society would be debating whether to send humans to nearby stars.

A majority of the twenty-nine respondents predicted a peaceful world, harmoniously unified under a democratic world government and freed from resource scarcity. Each entry had its optimistic idiosyncrasies. Edward Teller, one of the masterminds behind the hydrogen bomb, hoped that ballistic missiles would no longer be used to loft nuclear-tipped warheads, but would instead be turned to transporting passengers anywhere in the world in less than an hour. He doubted, however, that it would ever be "a comfortable way to travel." Vice President Lyndon B. Johnson opined that we might use satellites to control Earth's weather. The Republican California congressman James B. Utt thought that society would master the science and technology of human teleportation, though he didn't "look forward to this with any sense of enjoyment." Another California congressman, the Democrat George P. Miller, offered his curious opinion that by 2063 we would "have found humans living elsewhere in the universe besides Earth."

The strangest entry of all was the long, decidedly pessimistic response of Harold Urey, the Nobel-laureate chemist. Where few other contributors had filled more than a single page, Urey took up roughly a third of the entire book. His thoughts may well have been inspired by his participation in Frank Drake's Green Bank meeting two years earlier.

Urey hardly discussed space science and exploration at all. Instead, he devoted much of his essay to summarizing the social implications of changes he had seen in his lifetime. He had been a boy at the turn of the century, growing up when steam engines, railroads, telegraphs, and telephones represented the pinnacle of technology. He was now an old man, in a world filled with automobiles, airplanes, rockets, digital computers, color televisions, and atom bombs. He lamented how technological progress had cut off his children from many of the bucolic joys of his own upbringing, such as riding "in a sleigh behind a matched team of blacks, on a clear night with stars above and white snow around . . . nestled warm and cozy beneath a buffalo robe."

For every decade of Urey's life, human society had experienced a relatively constant factor of growth in technological capability and economic capacity. The expectation of continued growth—indeed, endless growth—was what underpinned capital investment in the ongoing research and development that was transforming the globe. But that exponential profusion, he wrote, was not opening up magnificent new frontiers so much as it was revealing previously unappreciated limits. Looking ahead, Urey glimpsed a not-too-distant future in which things could fall apart, when the centers of the modern world could not hold, a time when growth would stagnate. He postulated no proximate causes other than already-existing cracks in civilization's façade. Schemes for world government were unfavorable, he believed, because governments tended to grow bloated and cumbersome from "fantastic national debt" that outstripped both inflation and revenue. The ruinous deficits would be produced by "the curious psychology of politicians" paired with "the development of war machines by applied scientific methods," and would be exacerbated by the need to provide healthcare and social security for a large, aging populace. Turning society over entirely to the whims of large, private corporations was no alternative, Urey observed, because companies would inevitably conspire to pursue short-term profits against the public interest and common good. Only through some uneasy and uncertain balance between government regulation and private enterprise could the status quo of growth be maintained. Even then, it could not be maintained indefinitely.

Urey bemoaned the fact that most ordinary people were trapped in the present, scarcely able to consider a past predating the lives of their grandparents, unable to plan for a future beyond the lives of their grandchildren. Worse yet, he saw a public increasingly hostile to any scientific research and technological development that did not directly contribute to greater comfort and convenience in their daily lives. In developed nations, more effort than ever before was being channeled away from solving pressing global problems and toward making technological baubles to satisfy the cyclical desires of consumerism. Urey noted that U.S. fossil fuel consumption had increased eightfold between 1900 and 1955, much of it due to generating electricity. Further, usage of "electrical power increased from negligible in 1900 to about five hundred watts per person" by 1963. How long could energy usage continue increasing to support economic growth? In one of the few actual predictions to be found in Urey's response, he gently hinted such luxuries were unsustainable: well before 2063, he prophesied, we would be faced with the potentially unpleasant necessity of finding "ways to expend human energy other than by working on useful gadgets."

Energetic limits to economic growth are remarkably straightforward to calculate, given a few simplifying assumptions. Taking the United States as an example, data from the federal Energy Information Administration shows that the nation's total energy usage has grown by just under 3 percent per year since the middle of the seventeenth century. As a thought experiment, a UC San Diego professor, the physicist Tom Murphy, has calculated the consequences of that continued growth out into the future, extrapolating it to the entire globe and reducing it to 2.3 percent per year, which yields a factor-of-ten increase in energy usage every century. Starting from a circa 2012 global energy use of 12 terawatts, the world of 2112 would consume 120 terawatts, and the world of 2212 would consume 1,200. By 2287, world energy consumption would be 7,000 terawatts—an amount that could in theory be delivered by covering all the land on Earth with photovoltaic solar-power arrays operating at 20 percent efficiency. From there, increasing the efficiency of the photovoltaics to a miraculous 100 percent and covering the oceans as well as the continents would allow the 2.3 percent annual growth in energy use to persist for another 125 years, taking our steadily growing civilization into A.D. 2412 before it outpaced the total amount of sunlight falling upon the Earth. Another energy source, nuclear fusion, could potentially sustain an annual 2.3 percent growth rate for some centuries beyond this, at least until the waste heat from the vast amount of power being produced evaporated the oceans and turned Earth's crust to glowing slag. For a planet-bound civilization, the boiling point of water and the melting points of rock and metal place insurmountable limits upon the expansion of energy use.

Writing in the journal *Science* in 1960, the physicist Freeman Dyson carried humanity's recent, relentless energy consumption to its logical extreme, postulating that someday if we mastered living and working in space we could harness essentially all of the Sun's light by constructing a cloud of solar collectors around our central star. Dyson didn't sweat what he saw as relatively minor technical details, such as how we would acquire the vast amounts of necessary raw materials— he proposed that by the time we needed all the Sun's energy, we would be more than capable of simply dismantling a planet or two. Viewed from light-years away, the Sun's optical emission would fade and be re-placed by the infrared glow of waste heat emanating from its enclosing shell. If astronomers ever saw a distant star characteristically dim and shift entirely to infrared emissions, Dyson wrote, they would most likely be glimpsing evidence of another energy-hungry galactic civili-zation. Operating with perfect efficiency, such a "Dyson sphere" would capture some 400 billion petawatts of power—the total radiant output of the Sun. And yet, based on an ongoing 2.3 percent annual growth in energy usage, Murphy calculated it would cease to meet our expand-ing energy needs in just less than a millennium. There are, of course, a few hundred billion stars in the Milky Way. Assuming humanity somehow managed to instantaneously encase each and every Sun-like star in the Milky Way within perfectly efficient Dyson spheres, the in-exorable 2.3 percent increase in energy used per year would still bring us to the limits of our galactic capacity within another millennium.

"Thus in about 2,500 years from now, we would be using a large galaxy's worth of energy," Murphy has written. "We know in some de-tail what humans were doing 2,500 years ago. I think I can safely say that I know what we *won't* be doing 2,500 years hence." If technological civilizations like ours are common in the universe, the fact that we have yet to see stars or entire galaxies dimming before our eyes beneath starlight-absorbing veneers of Dyson spheres suggests that our own present era of exponential growth may be anomalous in comparison not only to our past, but also to our future.

Long before the General Dynamics capsule was entombed—at the onset of the Jurassic Period, to be precise—San Diego was ordinary marine limestone on the bottom of a seabed, much like the rest of what would eventually become present-day California. Sometime less than 200 million years ago, colliding tectonic plates caused vast batholithic plutons of magma—viscous city-size bubbles of molten granite—to surge up from the mantle into the crust beneath that ancient coastal ocean. The plutons were variously enriched with copper, lead, silver, gold, and other metals. They heated the waterlogged rocks from below, cooking limestone into marble. When magma mixed with water seeping down from above, some of the metals precipitated out to form veins of ore in overlying fissures. Over millions of years, the ongoing tectonic collision gradually thrust and uplifted the former seabed to become dry land. Great blocks of the crust were overturned, to lie across and bull-doze through the countryside in reverse stratigraphic order. A California mountain's summit might be made of granite from the subterranean depths, with sides formed from intermediate regions of ore-veined marble and limestone. Strewn along its base would be a jumbled melange of younger surface rock mixed with unconsolidated mudstone from the toppled ancient seafloor. Rain falling on the mountains eroded the sides, exposed the ore veins, and flushed flakes and fragments of precious metals into rivers.

On January 24, 1848, while building a sawmill along the American River to float logs to the small coastal settlement of San Francisco, a carpenter named James Marshall found a few pieces of that washed-down gold, sparking the great California Gold Rush. Soon, some 300,000 people from around the world had swarmed the region to seek their fortunes, exponentially increasing its population and propelling the unorganized territory into official U.S. statehood. Boomtowns bubbled and burst throughout northern California. San Francisco became a bustling city. The redwood forests fell to feed furnaces that reduced quarry-hewn limestone into lime, which went into the cement for marble-faced buildings. By 1863, a transcontinental railroad was under construction, and the great opening of the American West had properly begun. All because of gold, by chance delivered in a Jurassic upwelling of magma beneath the sea.

After the gold rush, the transcontinental railroad ensured that the surge of new settlers never truly abated. They rolled across the land in waves, chasing boom after boom, and at the end of each day, as the Sun fell into the Pacific, it set upon what appeared to be the truest expression of the American Dream. Almost everyone, it seemed, could make a fortune in the wide-open spaces of California. Farmers flocked to the Central Valley's mild climate and fertile soil. Oilmen found light, sweet crude locked away in the state's southern strata. Filmmakers found refuge in Hollywood from Thomas Edison's packs of patent lawyers back east. The U.S. military built bases, airfields, and shipyards along the Pacific frontier. Technologists gave birth to new high-tech industries in Silicon Valley. Throughout it all, real-estate speculators bought up parcels of land, subdivided and sold them, and became rich. Housing prices and infrastructural necessities rose as capital continued pouring in, and property taxes rose with them, until in the 1970s wealthy, established Californians rebelled. They voted to keep property taxes artificially low, and shifted the state toward a dysfunctional political culture where time and time again voter-led "ballot initiatives" earmarked spending while also eliminating sources of revenue. Since the turn of the millennium, the state had been in near-constant budgetary crisis. When the real-estate bubble burst in 2007, it helped kick off the Great Recession of 2008, which reduced California's coffers to catastrophic lows. Funding was slashed for public assistance to the poor and disabled, for state colleges and courts, for municipal emergency services, and more. For a time in 2009, the state government of California could only pay its debts with official, printed IOUs.

California's scars, old and new, were on display when I visited Laughlin at UC Santa Cruz. The campus is built over and around abandoned nineteenth-century limestone quarries and cattle pastures, surrounded by sparse redwood shadows of a great former forest. In a long sunlit hallway of the Interdisciplinary Sciences Building, which houses Laughlin's office, I came across a bulletin board filled with student notices protesting the school's budget cuts, flanked by two massive empty dewars meant to store liquid helium.

Laughlin no longer had the option of simply protesting the University of California's systematic belt-tightening. He was a tenured professor, and had just been appointed chair of the astronomy department. His office was small and sparsely decorated. A whiteboard, filled with knots of equations and hand-drawn graphs, took up much of one wall. A topographic globe of Mars rested on a file cabinet, surrounded by sparkling pieces of pyrite-laced granite Laughlin had fished from a nearby stream. In contrast to the globe's crimson-colored highlands and mountains, its basins and lowlands were tinted blue, like the seas they probably held billions of years ago. "No two days are the same; every day is exactly the same," Laughlin said as he showed me the office. "There's very little variation—I come in, I sit at my desk, and I work. But each day is a unique, mad scramble from one emergency to another to keep the research enterprise going. I'm trying to secure more funding for the department. I've got four grad students that all have to get funded. I'm making sure my own work is funded. Right now funding is like the old Martian seas—drying up at the margins."

Some of the younger members of his department, Laughlin thought, might soon leave the field, offering their analytic, numerical skills to more lucrative patrons in Silicon Valley or on Wall Street. In the meantime, some of his students and colleagues, hard-pressed for costly telescope time, simply made do with using cheap desktop computers to analyze free batches of public Kepler data, looking for valuable discoveries the Kepler team's initial analyses might have missed.

Laughlin himself was grappling with a mystery that American and Swiss RV surveys had glimpsed, and that Kepler had fully revealed: a wealth of Neptune-mass planets in hot, flat, circular, sub-100-day orbits, the apparent default architecture of inner planetary systems. Conventional theory holds that, since planets form from flat, circular, swirling disks of material around young stars, nearly all planets should reside in flat, circular orbits aligned to a star's equator, its ecliptic plane. Small, rocky planets would form close to a star, where it is hot and most gas evaporates away. Large, gassy planets would form farther away, past the "snow line," where gas and ice linger in the cold. Ever since their discovery in 1995, hot Jupiters had forced new theories. They were found in wildly elongated orbits, eccentric orbits that took them swooping far out of the ecliptic at one end, then plunging to graze their stars on the other. Theorists could only explain such worlds as products of planetary migration, a collection of theoretical mechanisms by which far-out giant planets could interact with their formative disks to bleed off momentum and fall closer to their star. The trouble was, when a massive planet began migrating mid-formation, it would tend to accumulate lots of material as it moved through the disk, growing to be about the size of Jupiter, not a relative runt like Neptune. Further, a migrating giant's gravitational influence would tend to shake up the rest of the system, scattering other planets from flat, circular orbits into elongated, eccentric paths tilted out of the ecliptic plane. According to most new theories, intermediate Neptune-mass planets simply should not exist close to their stars, and certainly not in flat, pristine orbits. Consensus held they could only form farther out, and that, migrating in, they would have grown larger while also disrupting the delicate coplanarity and circularity left over from the primordial process of planet formation.

Some theorists had been so confident that they predicted Kepler would find a "planetary desert" bereft of close-in Neptune-size planets.

And yet when Kepler's data began streaming in, it was filled with hundreds of transiting multiplanet systems—systems of hot Neptunes, some with orbits proportionally flatter and more circular than an LP or a compact disc. The predicted planetary desert proved to be a rainforest of inexplicable worlds. Theorists knew no migratory mechanism that could have transported the planets to their present positions with such quiescence. Yet there they were, like angry bulls pawing the dust amid floor-to-ceiling stacks of untouched, unbroken china. Finding one would have been a fluke; finding hundreds meant something fundamental was missing from the accepted theories of how planetary systems form and evolve.

Kepler's revelation had left Laughlin flabbergasted, like most every other player in the exoplanet game. He said as much as we left his office to walk among the redwood stands and old limestone quarries that laced the campus. The complete Kepler data, he believed, would keep theorists busy for the next twenty years.

"I just don't know yet how those planets formed," he said as we mucked along through a damp creek bed. "No one does. Something is wrong. Our paradigm of planet formation was built to weave two distinct things into a coherent picture: our own solar system, and the very easily studied hot Jupiters. We now know that hot Jupiters are only around maybe one percent of stars. Architectures like our own solar system seem to show up around more like ten percent or less of stars. So it looks like we've been trying to build a unified theory out of these two disconnected, fringe outcomes of planet formation. You don't need to be a scientist to see that's probably not the right way to go about it."

We climbed from the creek bed and began moving up a gentle redwood-covered grade. Laughlin likened the new planetary confusion to Bode's law, an idea named for and popularized by the German astronomer Johann Bode in 1772. Bode's law postulated that the spacing of planetary orbits followed a specific, harmonic pattern, one that neatly accounted for the observed orbits of Mercury, Venus, Earth, Mars, Jupiter, and Saturn. When Uranus was discovered in 1781, it fit the pattern too. But over the years, Bode's law fell into pseudoscientific disrepute as new discoveries such as Neptune, Pluto, and the asteroid belt failed to fit its harmony. Its early success had just been a coincidence, born of the fact that planetary orbits do indeed follow a much more general hierarchical spacing.

"The idea of a 'snow line,' with gassy planets only forming far from stars—that came about because it's what we see in our own solar system," Laughlin went on. "Now that we're seeing the reverse so often around other stars, I'm not sure anymore that the snow line is a relevant concept. It might be a modern example of Bode's law, but I don't think you're going to see people just stop talking about it. . . . You look at these new multiplanet systems, and in their orbital spacing and their masses in proportion to their host bodies, they follow the exact same patterns as the Jovian satellite system, Jupiter's big moons. And those moons probably formed right where they are now. I wouldn't be surprised if it turns out these worlds formed right where they are now, too. The process simply hasn't been studied very well because of the focus on our own solar system."

The redwoods thinned as we continued up the slope. Larger patches of midday sunlight blazed down through a canopy of ponderosa pines made sickly and sparse by bark-beetle infestation. We began to sweat. The earth's color had gradually shifted from grayish-brown at the creek bed to reddish-orange beneath the pines. A glitter in the soil caught Laughlin's eye, and he bent over to pick up a crumbly rust-colored rock. He apologized for his "grasshopper-level attention span," and pivoted to talking geology.

"This area, this rock right here, is transitional. It's the kind of rock you'd prospect for gold. It sparkles from the pyrite crystals. We've got limestone downslope that's been heavily cooked and altered with magmatic fluids. Upslope we've got the granite that did the cooking. The forest changes here, I think, because the redwoods like the limestone soil better. It's weird. The naive observation is that the granite formed where it lies and is somehow younger. You'd think as you go upslope you're reaching younger layers. That's what geologists thought for a long time. In reality, it's just the opposite. The theory of plate tectonics was a big leap that made more of the puzzle pieces fit together. We're going up in terms of topography but we're going down in stratigraphic layers because of how this entire block of crust has been tilted and eroded. The older rock is on top of the younger rock here. We could walk this long, slow grade for another fifteen kilometers, and the whole way up it's granitic plutonic rock from deeper and deeper down."

I asked him how he knew all this. He said, "My sense is, if you really want to understand Earth-like planets, you have to become an expert on the Earth."

Back in his office, Laughlin elaborated that a good bit of his knowledge of Santa Cruz's geology actually came from a lingering fascination with financial markets. He gestured to his office's equation-filled whiteboard. That particular day, the differential equations didn't directly concern astronomy at all, he confessed, but rather the fluctuating commodity prices of precious metals, which he wished to predict on timescales of months and years. To do that, he had needed to understand supply and demand—the cost of building new mines, the manner in which metals were extracted and used—so he had educated himself in the geology of ore-forming bodies, the same geology that so long ago put gold in California's hills.

Laughlin had reveled in using his skills in a technical field with direct commercial applications. Astronomy was a technical field, he acknowledged, but for as long as humans remained a single-planet, single-star species, it would be disconnected from the profits associated with semiconductor physics, petroleum prospecting, or quantitative finance. His new investigations had borne strange fruit: he had become preoccupied with what certain obscure market trends revealed about the nature of prediction and the formation of monetary value, and had begun to closely scrutinize various trades as they propagated through the global financial system. Viewing the scintillating patterns of trades, his "bird's-eye view" on what would otherwise be "a mostly hidden world," Laughlin had begun feeling the familiar effervescence once again. He saw battle lines being drawn in virtual space, occasionally spilling out into the real through volatile bubbles of liquidity, low-latency arms races, and high-frequency information warfare. The vista newly unveiled before his eyes was of a planet on the brink of some profound transformation, one driven as much by high-speed telecommunications and computing as by biology and geology.

In its reliance on the high technologies of ultrafast mainframes, undersea fiber-optic cables, microwave relay networks, and communications satellites, the frontier of modern finance almost seemed to constitute a new Space Age, though one that would be unrecognizable to the astronauts and rocketeers of half a century before. The planet's brightest scientific minds no longer leveraged the most powerful technologies to grow and expand human influence far out beyond Earth, but to sublime and compress our small, isolated world into an even more infinitesimal, less substantial state. As he described to me the dark arts of reaping billion-dollar profits from sub-cent-scale price changes rippling at near light-speed around the globe, Laughlin shook his head in quiet awe. Such feats, he said, were "much more difficult than finding an Earth-like exoplanet."

In early January of 1848, a fifty-two-year-old master carpenter named James Lick arrived in the small village of San Francisco. He had been born in Pennsylvania but had made his fortune building and selling fine pianos in South America. He hoped to expand that fortune through purchasing cheap land in the new California territories, which he thought would soon be annexed by the United States. Along with his tools and workbench, Lick had brought along an ironclad chest filled with $30,000 in gold. He immediately began buying up vacant lots around town. Seventeen days after Lick's arrival, James Marshall discovered gold at Sutter's Mill, the California Gold Rush was set in motion, and Lick found himself the biggest player in a buyer's market for San Francisco's abundant real estate. Soon he was swamped with sales offers, as residents abandoned their coastal harbor homes in droves to seek gold in the inland hills. He bought up all the land he could at cut-rate prices, then netted huge profits as San Francisco's population exponentially boomed from wave after wave of arriving prospectors. Within a decade he had become one of the new state's most successful land barons, with vast holdings in San Francisco, Santa Clara, and San Jose.

By the time Lick suffered a debilitating stroke late one 1874 evening in the kitchen of his Santa Clara home, he was the richest man in California. He spent his remaining years in convalescence, planning the fate of his fortune for whenever he died. Lick's first thought was to build giant statues of himself and his parents, so gargantuan they could be viewed from far out at sea. He was dissuaded when he realized their visibility would make the statues prime targets for any future naval bombardment of the coast. For a time he wished to build a pyramid larger than any in Egypt on a large tract of land he owned in San Francisco. Lick changed his mind again, thanks to the persuasiveness of his friend George Davidson, an astronomer and president of the California Academy of Sciences, who convinced Lick that he should instead build the world's most powerful telescope. Lick ultimately drew up a deed of trust that bequeathed $3 million to various public works throughout the state that had made him so wealthy. Seven hundred thousand dollars went to the University of California to be spent building an astronomical observatory to house "a telescope superior to and more powerful than any telescope yet made."

A few months before his death in 1876, Lick signed off on his trustees' choice of Mount Hamilton as the site of the observatory, and told them he wanted to eventually be buried beneath his great telescope. Construction began in 1880, and the 57-foot-long, 36-inch "Great Lick Refractor" achieved first light in 1888. It was enclosed within "the first hot-rivet project west of the Mississippi," a beautiful neoclassical dome of green-patinated metal. Lick's refractor was the world's most powerful telescope for nearly a decade, and to this day remains the second-largest refracting telescope on Earth. After the 36-inch's completion, in accordance with his wishes Lick's remains were disinterred and entombed below the observing floor of his telescope. There, beneath curving beams that resemble a fine piano's wood-carved hammers, a spotlight shines down through the gloom onto fresh flowers and a brass plaque that reads "Here Lies the Body of James Lick."

Before the Great Lick Refractor's completion, Lick's trustees had inaugurated the observatory by building a small, dome-sheltered 12-inch telescope. On December 6, 1882, the 12-inch was used to observe the transit of Venus, along with another telescope custom-built for the occasion. The astronomer David Peck Todd traveled from Massachusetts solely to use Lick's equipment to observe the transit, and was fortunate to arrive beneath clear skies. Over the course of four hours, he captured Venus crossing the Sun on 147 chemically treated glass plates, creating the most complete photographic record of a transit prior to the twenty-first century. It would be another 122 years before Venus once again cast its shadow down upon Mount Hamilton.

The sky was gray and ominous above Lick Observatory when I arrived with Laughlin on the afternoon of the appointed day—June 5, 2012. In the intervening years, Mount Hamilton had sprouted about ten more major telescopes, all in white domes sprinkled across the summit. Marcy and Butler's Automated Planet Finder sat atop a nearby crag, soon to be unleashed upon the sky. Behind it loomed the enormous dome of the 120-inch Shane reflecting telescope, the largest on the mountain and the primary eye on the sky through which Marcy and Butler had discovered most of the first hundred worlds in the early years of the exoplanet boom. The Hamilton spectrometer they had used for their first RV surveys was still in the Shane's basement, though upon decommissioning it would be moved to the Smithsonian Institution for display and safekeeping as a national treasure. As with all observatories near large urban areas, Lick's utility had suffered in recent decades from electric light pollution. Its astronomers lived for nights when Pacific fog rolled in to blanket Silicon Valley below. The coastal city lights would vanish and the stars would shine down like diamonds from a sky as dark as it had ever been in all the eons before. Perched on its peak above the fog, the aging observatory still had life left in it, somewhere between a few years and forever. Another transit of Venus would not occur until December 11, 2117.

Laughlin and I entered the small dome that once held Lick's first telescope, the 12-inch. The dome now contained the squat, stubby, rust-colored Nickel Telescope, built in 1972. Its 40-inch mirror was larger and more powerful than the lens of the Great Lick Refractor of yore, but still so vastly inadequate compared to the best telescopes of 2012 that it was rarely used for cutting-edge observations. That afternoon, however, it was poised to capture the Venusian transit in a literal new light. Sloane Wiktorowicz, a trim and athletic thirty-year-old UC Berkeley postdoc, had mounted his custom-built instrument, POLISH, on the Nickel for the occasion. He was folded into a chair in the small adjacent control room, monitoring his instrument using three large flat-screen displays. Not every element of the endeavor was high-tech: POLISH was shrouded from ambient light by a black nylon sheet held in place by duct tape and cardboard from a disassembled box of Famous Amos Chocolate Chip cookies. The Nickel's mirror would have melted beneath the Sun's concentrated rays, and was mostly shielded by a block of wood. A small hole drilled in the wood and covered with a silver filter let sunlight safely trickle in. Lashed with more duct tape to the telescope's frame was a sawed-off section of black-painted stovepipe. It extended above the mirror like the barrel of a gun to further minimize any light from within the dome reaching POLISH's delicate sensors.

POLISH measured polarization, the way that the waves in a beam of light oscillated at perpendicular angles to the light's direction of travel. Light ordinarily is unpolarized—meaning each photon's polarization oscillates in a random direction—but when it reflects or scatters off a surface, or an atmosphere, the light can become polarized, with each photon's oscillation aligning in the same direction, like iron filings in a magnetic field. The principle is used in polarized sunglasses, which reduce glare by filtering out polarized light reflected off clouds in the sky or the surface of a lake. In astronomy, Wiktorowicz told me, the same effect could be used to enhance the signal of polarized light bouncing off or shining through an exoplanet's atmosphere. Measure that light's polarization very carefully and you could get data about clouds, hazes, and atmospheric composition for a planet, even one that was many light-years away. Forty years earlier, Wiktorowicz said, polarization measurements of Venus had provided the first evidence that droplets in its atmosphere weren't made of water but of sulfuric acid. His task on this day was to observe Venus just as it started its transit, during ingress, when it blocked part of the Sun's edge and exhibited the strongest polarization signal. Out of the transit's six-hour duration, Wiktorowicz had a fifteen-minute window at the beginning in which he could make his measurements.

"The idea is to calibrate what we can expect from exoplanet transits," Wiktorowicz explained. "We're seeing Venus from up close at a distance of only about a third of an AU, which means its angular size is about three times larger than if we were seeing it from outside the solar system, like you'd see an exoplanet. The interplay between those two scales means that when we look at this transit from Earth, it's as if we are witnessing one of these hot Neptunes we see in the Kepler data transiting the Sun, in terms of the total amount of starlight that gets blocked and the size of the atmospheric ring. I don't know if anyone else will be looking at the transit in polarized light. This is a once-in-a-lifetime opportunity."

Twenty minutes before the transit's beginning, pools of blue were breaking through the cloud banks above Lick, but the weather was still uncertain. Wiktorowicz had bigger problems. The computer in charge of guiding the 40-inch telescope had suddenly gone berserk, sending the telescope, POLISH, and duct-taped improvisations slewing wildly, perilously close to colliding with the concrete floor, the metal dome, or anyone foolish enough to stand in the way. He couldn't even point the telescope, let alone take polarization measurements. Pavl Zachary, a scruffy technician wearing Army-standard olive drab, stormed into the small control room, out of breath, a squawking walkie-talkie hanging from his belt. "I was just settling in for the transit with a bag of celery and some cheese puffs," he said between gasps. "Then I get a call on the radio that Sloane is trying to do science. The nerve of some people! Sloane, are we parked?"

"Supposedly the telescope is parked," Wiktorowicz replied, looking at the Nickel, which had begun slewing toward the floor, as if to gain a view of Earth's core, or of Madagascar on the opposite side of the world. The transit of Venus would begin in ten minutes.

Zachary began to do battle with the telescope's computer, resetting its various components one by one, occasionally clambering up into the dome to watch for results.

"This is bad luck, but it could be worse," Laughlin joked, trying to lift Wiktorowicz's flagging spirits. "At least you're not Le Gentil! You don't have dysentery, you still have all your possessions. You haven't been declared legally dead."

"There is that," Wiktorowicz said with a mordant chuckle. "I don't have dysentery and I'm not dead yet. I'm guessing I will be by 2117."

"It's bypassed the software limits—that's kind of scary," Zachary eventually said. "I've never seen anything remotely like this before. Our robotic telescope can't even find the Sun. We might have to search through the incident reports. It's times like this I'm glad we do not work at a nuclear power plant." He scratched his head. "Sloane, are you ready for the really bad news?"

"Sure."

"The associate director is on the mountain for the transit."

"Oh. Oh, no." Wiktorowicz glanced over at me, then explained: "Things tend to go wrong when he's around. He's a really nice guy but seems to have bad karma at observatories."

"Terrible karma," Zachary chimed in from the dome, his voice muffled by the Nickel's whirring hydraulics.

Wiktorowicz nervously drummed his fingers on his desk in time with beeps from the Nickel's motion alarm. He looked at his open net-book nearby, tuned to a NASA video feed from the Keck telescopes in Mauna Kea, Hawaii. The picture switched from three commentators bundled against the mountain chill to a telescopic view of the entire Sun, reddened by a filter. A small black arc appeared and slowly grew at the Sun's edge, like a worm's bite out of an apple. Ingress had begun. Minutes passed. Muscles worked and writhed along Wiktorowicz's jaw, and a bead of sweat slid down his temple in the cool air of the control room. Blue sky could be seen through the dome's open slit, above the aimlessly drifting telescope. The clouds had cleared. He sighed, cursed, produced a turkey sandwich from his bag, and ate it with resignation.

"This seems excessive, even for the associate director," Wiktorowicz said between bites. "It's like the telescope just lost its mind. Maybe the ghost of that French guy with dysentery is trying to stick it to us."

"I think it just drifted into a rough mood and needed to cool off with a random walk," Zachary said. "Cheer up, Sloane. We're gonna help it find itself."

Laughlin and I excused ourselves to go watch the transit from the parking lot beneath the suddenly clear sky. As we left I glanced again at the netbook's video feed from Mauna Kea. Wiktorowicz was staring dejectedly at the screen, slowly chewing another turkey sandwich. The worm-bite arc had become a perfectly circular bullet hole in the Sun— Venus had slid well within the disk, and its ingress had passed.

We walked out into bright sunlight, seemingly undiminished by Venus's shadow. I risked a quick, dazzled glance up at the glaring Sun, but it looked as it always does. Cumulus clouds still dotted the sky, and when one every now and then drifted over the Sun, a soft wail would rise up from the gathered crowd, followed later by cheers when an un-obstructed view returned. It would be another few hours before the Earth's rotation carried the Sun and the transit's conclusion over the horizon and beyond sight. Wiktorowicz eventually admitted defeat and left the confines of the Nickel dome to join us. Through one of the small telescopes nearby, we took turns gazing at Venus's imperceptibly creeping black circle and clusters of nearby sunspots. Few words were said. The silence deepened with the acceptance that each gaze brought the experience closer to an end, and that in all our lives we would never see such a sight again. When the Sun had sunk low in the sky and chill winds were rising at the edge of twilight, Laughlin said his goodbyes, and we began the drive back to Santa Cruz.

Weaving down the long, winding road, Laughlin remarked that the transit had surprised him. He had thought it would be more like a total solar eclipse he had witnessed years earlier.

"I was in a boat off Baja, Mexico. It was July, but about ten minutes prior to totality the air started to get noticeably colder, seemingly every second, as the Moon blocked off more and more sunlight. The Sun visibly became a crescent, and the optical effects were overwhelming. Everything seemed to be swimming, the shadows were all distorting into little crescents, and the light was becoming very sharp and angular. I looked up and saw shadow bands flowing overhead as the light shined through convection cells in the upper atmosphere. I looked down and saw the eclipse's shadow sweeping across the ocean toward me at breathtaking speed. Then the Moon slid into place, and sunlight shining through its mountains and valleys drew a diamond ring in the sky. The Sun's corona popped out, white and glowing and wavering. I could see the planets all stretched out along the ecliptic—Mercury, Venus, Mars, Jupiter. The whole solar system was right in front of my eyes. Orion was directly overhead. Everyone was hooting and hollering and yelling. It was pure primal joy, like that feeling right after a big football touchdown. The eclipse itself lasted something like seven minutes, but it went by in a flash. There was no time for contemplation or anything deep. It was a roller-coaster ride.

"The transit of Venus couldn't have been more different," he went on. "Right as it began everyone was running and rushing around, stumbling over themselves to see it and exclaim how wonderful it was. But then it just went on for hours. People got quiet. It gave everyone time to reflect and more meaningfully sense for just a short while some small part of this much greater, grander picture of our place in things."

I asked Laughlin what he had reflected on during his long silence on the mountain.

"The transit," he began. "That period when a terrestrial planet hangs in front of its star. It reminded me that there is only a fleeting moment in the universe's entire history in which it can have planets with water on their surfaces around a star like the Sun. And here we find ourselves in its midst, at this profound boundary in the history of the Earth, in this instant of time where a major geological age is ending and another one of our own making is beginning. There's no way to predict what exactly will happen during the transition, but I don't think we'll just fade away. We have yet to make our maximum impact, whatever it may be."

CHAPTER 6

The Big Picture

When the alarm woke Mike Arthur at a quarter to six, he rubbed the sleep from his eyes and shuffled into the kitchen of his two-story white farmhouse to put a pot of coffee on. As it brewed, Arthur went to a window and looked out at the surrounding valley, surveying the twenty-six acres of pasture and woodland that he tended with his wife, Janice, and their two college-age daughters. It was dark in the glen, more than an hour until the first rays of sunlight would crest the larch-lined hilltops. Only the murmur of a nearby creek broke the thick arboreal hush that hung around the farm. In the glen's early-morning serenity, uncorrupted by the sound of cars or planes, televisions or radios, it seemed for a moment not to be late October, 2011, but long ago, in a time before clocks, before calendars. Before man.

The coffeemaker beeped its completion, and Arthur turned from the window to pour a cup, which he left cooling on the counter. He stepped into overalls, then pulled on a jacket over a chest and shoulders broadened by a youth spent surfing in California and preserved by an adulthood of farming in central Pennsylvania. When he stepped outside, Arthur was surprised to find the autumnal air so warm he couldn't even see his breath.

Trailed by his sheepdogs, Arthur walked to the barn to give water to his flock of Icelandic sheep and change their hay. Next, he fed the free-range chickens and gathered their eggs, then entered a small greenhouse to check his crops of organic kale and chard and his seedlings of seasonal vegetables. He adjusted the vents to allow more of the cooler outside air to flow in—by the afternoon, it would be another unseasonably balmy day, and he didn't want the plants to bake.

Back inside, Arthur showered, then turned to the bathroom mirror and spread shaving cream across his cheeks and beneath his jaw, up against the edges of a full-grown white goatee that tapered to a point at his chin and swept out into prominent whiskers beneath his nose. His face was rugged, reddened, and lined from years of exposure, and his high forehead was topped by a peak of long gray hair pulled back into a ponytail. All that, plus his slight paunch and burly frame, made him look a bit like Saint Nick, by way of Santa Monica. As he shaved, Arthur thought back to two decades ago, when he and Janice had begun their farm. Had the growing seasons and temperatures really changed since then? There could be no doubt about it: most passing springs and autumns now seemed more fleeting, faded into longer, hotter summers and milder, shorter winters. Next year, he thought, maybe he'd risk moving some of the more cold-tolerant vegetables out of the greenhouse to clear room for delicate out-of-season produce, which he could sell for a premium price at nearby farmers' markets. He splashed water across his face, dressed, gulped his coffee, kissed Janice goodbye, and hopped in his car to drive twenty miles west to his day job in the town of State College, at Pennsylvania State University, where he was a professor of geology.

Mike Arthur was a sedimentary geologist. Viewing walls of rock with alternating bands of limestone, sandstone, shale, and coal was for him like reading stories, ones written in stone. He was also a geochemist. With the help of a hammer, a sample bag, and a bit of laboratory wizardry, he could discern the subtle chemical signals in rock layers that revealed ancient, long-vanished environments—the flora and fauna, the weather and geography, and how each former world developed, flourished, and finally passed away, largely forgotten but for those lithic memories.

Paleoclimates and past global climate changes were his specialty, as seen through his research emphasis, the formation of black shales. Black shales are compactions of clay, mud, and silt formed in deep water and made the color of jet by their heavy loads of organic carbon. Organic carbon—the stuff from which plants and animals are made— is normally quickly eaten and recycled in a water column. But when organic detritus drifts to the stagnant bottom of a deep body of water, the absence of sunlight and oxygen can stave off the creatures that would otherwise churn through and consume the remains. Undisturbed, layers of carbon-laden silt and mud accumulate, compress, and sink deeper beneath the Earth's surface, where a slow geothermal simmer cooks them into black shale. Given sufficient heat, pressure, and time, a fraction of the carbon in organic-rich black shale transforms into petroleum, and further cooking will crack the oil into methane and a handful of other volatile organic compounds collectively and colloquially known as natural gas. To Arthur, instances of worldwide black shale deposition were signposts of past pulses of global warming: as temperatures climbed and sea levels rose, the deepening, tepid oceans would have lost much of their ability to mix oxygen-rich surface water to the bottom. Anoxia would set in, and rich deep-sea ecosystems would dissolve into sulfurous, bacteria-infused black mud.

Arthur's research into black shales initially took him around the country and the world, but by the early 1990s he had decided to settle in Pennsylvania. There, he realized, a good bit of the crucial evidence he needed to study black shales—and the Earth's fluctuating climate over the past 500 million years or so—could essentially be found right in his backyard, in the Allegheny Plateau. The Allegheny Plateau boasts some of the world's largest black shale deposits. In their finest details, the shales and their surrounding rocks told of the comings and goings of mountain ranges, glaciers, and vast inland seas in Pennsylvania's deep past.

Pennsylvania's rocks are also intimately linked to our planet's climatic present and its future. The inexorably rising temperatures—temperatures that were sending glaciers and polar ice into retreat, strengthening storms, shifting animal migration patterns, and making Arthur reconsider his greenhouse seedlings—in a way had come from the very ground beneath his feet. The additional warmth came chiefly from rising levels of atmospheric carbon dioxide, CO_2, a gas prodigiously produced by the combustion of fossil fuels. Carbon dioxide is transparent to visible light but absorbs a good fraction of infrared light—that is, light we perceive as radiant thermal heat. Sunlight readily passes through the gas on its way to shine on Earth's surface, but when the warmed surface re-radiates that light skyward in the infrared, it becomes trapped by the absorptive blanket of CO_2. This is the basis of the well-known "greenhouse effect," and CO_2's greenhouse effect is believed to be the primary architect of Earth's climate now and for the last half billion years. Humans had been gradually raising the atmospheric levels of CO_2 and other greenhouse gases for thousands of years, mostly through agriculture, but the rate of increase had greatly accelerated in the industrialized boom times of the past century. Much of that sudden surge had its roots in the rocks of the Allegheny Plateau, which runs through Pennsylvania and into portions of surrounding states.

The Big Picture

• • •

The largest known anthracite coal deposit on Earth was discovered in northeastern Pennsylvania in the latter half of the eighteenth century, supposedly when a hunter building a campfire accidentally set a nearby outcropping of crystalline black rock ablaze. By the mid-1800s, Pennsylvania anthracite had supplanted wood as the preferred method for heating homes in the United States, and coal mining had become a major industry throughout the Allegheny. At about the same time, Pennsylvania gave birth to the global petroleum industry, when drillers of salt wells found their work hampered by thick, viscous upwellings of black "rock oil." The first petroleum refinery was built in Pittsburgh in 1853, and the first oil well in the United States was drilled near Titusville, Pennsylvania, in 1859. Petroleum found its killer app in Henry Ford's Model T, which first rolled off a Michigan assembly line in 1908. The U.S. natural gas industry was actually birthed just north of the Pennsylvania state line, with a well drilled in Fredonia, New York, but the black shale deposit from which it came proved to have its bulk in Pennsylvania territory.

Riding on the surge of ancient carbon, Pennsylvania's economy boomed. Oil wells and mine shafts soon suffused the Allegheny rock, and refineries, pipelines, and railroads sprouted like weeds across the state. Like most booms, this one was short-lived. Output from the state's oil fields had already begun to decline by the dawn of the twentieth century, and was progressively overshadowed by immense newly discovered fields in Texas, Venezuela, Saudi Arabia, the Gulf of Mexico, and elsewhere. By the 1950s, Pennsylvania's Allegheny rocks still contained abundant coal and gas, but in a world increasingly addicted to oil, market forces dictated that those less-profitable fuels simply be left in the ground.

Pennsylvania's energy fortunes sharply rebounded in the first decade of the new millennium. As oil production from conventional, easily accessible reservoirs peaked, energy companies devised new methods to wring more oil and gas from harder-to-reach, "unconventional" source rocks. The most successful new method was hydraulic fracturing, or fracking, which squeezed previously inaccessible natural gas from deeply buried shales. When a gas-bearing shale lies beneath miles of rock, as it does throughout the Allegheny, the resulting pressure can lock gas within the formation. Pumping millions of gallons of high-pressure, chemical-laced water down a borehole, however, splinters the shale rock, and granules of sand or ceramic added to the slurry prop open the fractures. The locked-in gas, now liberated, streams through the cracks and back up the borehole, to be collected, compressed, and sold.

Fracking, combined with technology for drilling wells not only down but also laterally across layers of rock, offered a way to tap the biggest black shale formation in the Allegheny: the Marcellus. It was named for a small town in upstate New York where it jutted from the ground in sheer, flaky cliffs of carbon, and its expanse stretched westward from New York's Finger Lakes to the eastern half of Ohio, and south down to Maryland and West Virginia. But the Marcellus's concentrated carbonic heart could be found a mile or more beneath most of Pennsylvania, conveniently abutting major, energy-hungry metropolitan areas across the northeast United States. Comparing production rates of Marcellus fracking operations with the deposit's extent, thickness, depth of burial, and shale porosity, one of Mike Arthur's Penn State colleagues, the geologist Terry Engelder, estimated the formation might hold nearly 500 trillion cubic feet of recoverable gas. That would be enough to make the Marcellus the second-largest known gas field on Earth, enough to supply the entirety of the United State's energy needs for two decades.

As word spread of Engelder's Marcellus evaluations, energy companies great and small swooped in, buying up leases by the truckload in rural communities. A new boom began. Some farmers with huge tracts of land overlying productive parts of the Marcellus became millionaires overnight. Restaurants, motels, and other businesses sprang up to meet the needs of an inrushing flood of new workers. But the boom had a dark side, too. Long stretches of backwoods roadway buckled beneath roughshod convoys of heavy trucks, and sylvan forest glades disappeared beneath parking-lot-size concrete drill pads and miles of snaking pipeline. Natural gas linked to nearby fracking operations found its way into well water, and concerns grew about the possibility of fracking's proprietary chemical cocktails contaminating regional lakes, rivers, and aquifers. Public opposition soared, particularly in major cities served by the vulnerable watersheds. Penn State, keenly aware of its long and lucrative association with the oil and gas industry, attempted to walk the line between opposition and support. It formed the Marcellus Center for Outreach and Research in 2010 to engage with and educate all the region's stakeholders on the pros and cons of further developing the shale. The university chose Mike Arthur to serve as co-director of its new center.

On that unseasonably warm October day in 2011, a few hours after he had arrived from his farm, Arthur and I sat in his fifth-floor office talking about the Marcellus. He pulled up an animated time-lapse map on his desktop computer to show me the year-by-year progression of Marcellus shale drilling in Pennsylvania. The deposit's expanse was indicated by the color yellow, which filled most of the state, with a dot for each new well. Sixty dots sprinkled the yellow state map for 2007, the year of Engelder's initial estimate. In 2008, the number of new wells jumped to 229. Six hundred eighty-five wells were drilled in 2009, followed by another 1,395 in 2010. Nineteen hundred and twenty more had come online in 2011. On Arthur's computer screen, yellow, pockmarked Pennsylvania looked like a slice of Swiss cheese.

I asked Arthur to give me the gist of how all that energy, all that carbon, had found its way a mile and a half below Pennsylvania. He gestured to the map, to the south central portion of the state, where an arc of gray, wrinkled land crested above the surrounding yellow. There were no drill dots on those gray folds, because there was little shale beneath them. They were the Allegheny Mountains, a northern offshoot of the vast Appalachian Range. Geologists believe they peaked approximately 290 million years ago, in a mountain-building event called the Allegheny orogeny, one tiny event in the motions of Earth's tectonic plates that gradually thrust Europe, Asia, and Africa all against what is now North America to form the supercontinent of Pangaea. The Alleghenies had likely once been at least as tall as the Rockies or the Alps, or even the Himalayas, before being worn into gentle, rolling ranges by hundreds of millions of years of wind and rain. Beneath the Allegheny surface folds, Arthur said, there were layers of debris from a succession of more ancient, eroded ranges, each linked with its own pulse of mountain building and tectonic collision. One of those pulses, associated with the Acadian orogeny nearly 400 million years ago, in the midst of a span of time we know as the Devonian Period, was what had set the stage for the Marcellus.

The world was warm during much of the middle Devonian, too warm for polar ice caps. Some of the water that would otherwise have been locked up as ice was instead thinly spread over the North American interior as a shallow inland sea. Most of what is now Pennsylvania was flat, and underwater. Meandering continental drift had yet to transport it to its present northerly locale—at the time, it was in tropical latitudes. Phytoplankton, fish, and squid-like nautiloids thrived amid coral reefs and sponges in the warm, clear seawater. In death, their calcareous bodies, skeletons, and shells came to rest in thick layers of white lime mud on the seafloor dozens of feet below. The remains gradually hardened into rock layers of calcium carbonate—limestone. Eastward was an ocean, though not the Atlantic. It was the Paleo-Tethys, and it was disappearing, squeezing shut between landmasses on geological collision courses. Island arcs appeared on the eastern horizon, harbingers of the Acadian orogeny, foot soldiers on the front line of a tectonic advance. Over tens of millions of years, the island arcs approached and collided with the continent, slowly lifting mountains on the land like folds rising in a rug pushed across a slippery tile floor. Ranges took root in what would later become New York, New Jersey, Massachusetts,

Delaware, New Hampshire, Maryland, and south central Pennsylvania. Pressed down by the weight of the surrounding mountains, the crust—the planar lime-mud floor of the sea—subsided and sank perhaps 200 meters (700 feet), centimeters per millennium, carrying the seafloor ecosystem down to destruction, far below the penetrative power of life-giving sunlight. Algae, phytoplankton, and the rare fish were all that was left behind in the open surface waters. In the dark depths of that sunken inland sea, the Marcellus shale was born.

"Picture this sea, surrounded by mountains at least a mile high, largely cut off from the world ocean," Arthur said. "The mountains made their own weather, and then slowly weathered away. It's called an orographic effect. They lifted up air masses and formed storms that rained out over the peaks. Erosion carried huge volumes of sediment and nutrients into the water. Iron, copper, zinc, phosphorus, molybdenum. The nutrient influx really ramped up the productivity of the algae and the phytoplankton, which bloomed, died, and decomposed on the seafloor. The decomposition used up a lot of oxygen, more than could be replaced by turnover and circulation in the deep water. That was great news for anaerobic, sulfate-reducing bacteria already living on the bottom. Oxygen is toxic to them; they are some of the planet's most ancient organisms, from before our atmosphere had abundant oxygen. Anyway, they release hydrogen sulfide, which is toxic to most everything else. So those bacteria really knocked out whatever benthic ecosystem was left. After that, whatever organic matter settled to the bottom had practically nothing to decompose and recycle its carbon. The environment shaped the bugs, and the bugs in turn shaped the environment. That coevolution was what made the Marcellus."

Over the course of about two million years, a fine particulate rain—countless trillions of little deaths—continually drifted down to the anoxic bottom, forming layer after layer of pristine organic carbon. At last, the underlying crust accommodated the weight of the mountains, found equilibrium, and stopped subsiding. Sediments continued to course in from the eroding ranges, piling on over the thick black mud, burying an entire sea's worth of carbon. Eventually they piled so high the seafloor was raised once again into sunlight, and oxygen-rich, clear-water ecosystems returned—but only for a short while. Almost entirely filled with accumulating sediments, and now fully cut off from the global ocean, the vast basin's last vestige of sea gradually evaporated. Millions more years passed, and the mountains wore down to stubs, burying what would become the Marcellus even deeper beneath their scattered strata.

Removed from its ancient context, the creation of the Marcellus struck me as eerily familiar. A new source of energy and nutrients flows into an isolated population. The population balloons and blindly grows, occasionally crashing when it surpasses the carrying capacity of its environment. The modern drill rigs shattering stone to harvest carbon from boom-and-bust waves of ancient death suddenly seemed like echoes, portents of history repeating itself on the largest of scales.

And yet, as grand as the changes were that created the Marcellus—the collisions of continents, the rise and fall of mountains, the burial of an entire sea—they paled in comparison to an even greater global transformation that began at approximately the same time, Arthur explained. The Marcellus was the last major black shale that contained no significant debris from land plants, he said. When the mountains rose around that nameless sea, they were likely bald, and the rivers that washed down from their steep slopes flowed in roaring braids through a landscape devoid of vegetation other than scattered mosses, lichens, and fungi. At that point some 390 million years ago, a point seemingly so far removed from the present day, the planet was already well over four billion years old. And in all of that time not even a single green leaf had graced the entire terrestrial world.

"This was a time of transition, when vascular plants were just beginning to colonize the land," Arthur told me. "They start cropping up in black shales just above the Marcellus, and as the shales get younger you start seeing more and more evidence of land plants. You get into the late Devonian rocks, and you can see fossilized land plants that seem to first be colonizing around riverbanks and shorelines. They had yet to fully invade other life zones farther from the water. It's kinda cool."

Two evolutionary innovations spurred the colonization of land, each involving the harvest and transport of water. Land plants "vascularized," developing roots to draw water and nutrients from the earth, and they also began building their bodies from lignin, a durable carbon-rich macromolecule strong enough to bear water's heavy weight. The resulting vascular, lignin-rich plants propagated across the continents. They doubled the planet's photosynthetic productivity and dramatically altered the planet's carbon cycle. Once again, life and its environment were shaping each other in a powerful, world-changing feedback loop.

In death, the durable lignin in the leaves, stems, trunks, and roots of the new land plants resisted easy decay. When submerged by floods and sedimentation, all that vegetal carbon became locked away for hundreds of millions of years. Over time the plant remains turned to peat, then lignite, and finally coal as their depth and duration of burial increased. The process peaked in a 60-million-year geological period that followed the Devonian, when so much lignin-locked carbon was buried and converted to coal that it formed massive deposits around the globe, including the high-grade anthracite and great coal measures of Pennsylvania and the surrounding Appalachian states. Geologists appropriately call this time the Carboniferous Period.

Back in the late Devonian, oxygen that would have otherwise bonded with carbon decomposing in the open air instead built up in the atmosphere, probably reaching concentrations nearly double that of the present day. This rise in atmospheric oxygen coincided with the first insects and amphibians leaving their aquatic environments to fly, crawl, and walk the Earth. In Pennsylvania and elsewhere, their fossilized remains are often found in late-Devonian "red beds," deposits of iron-rich sedimentary rock that rusted when they were saturated with atmospheric oxygen. The high oxygen levels and new abundant fuel from land plants also increased the frequency and severity of wildfires, which may have prompted the evolutionary shift from fragile spores to hardier seeds that could endure high-temperature, low-moisture conditions during and after a conflagration. The emergence of seeds allowed plants to propagate from the moist coasts and lowlands into drier highland environments. For the first time in Earth's history, mountains and continental interiors were blanketed in green.

The rise of vascular land plants caused so much carbon sequestration during the late Devonian and early Carboniferous that atmospheric CO_2 levels plummeted. The diminished greenhouse effect dropped global temperatures by only a few degrees, but that seemingly slight change was enough to tip the world into a long-term ice age. Ice caps formed and grew at the poles as cooler summers failed to melt the accumulated snows of previous winters. The bright white spreading glaciers reflected more sunlight into space than the darker lands and seas, driving temperatures lower still. On average, every few tens of thousands of years or so, glaciers advanced from the poles into lower latitudes, locking water in their icy clutches to reduce global sea levels and turn climates more arid. Each time, terrestrial species in polar and temperate latitudes were forced down to the tropics ahead of advancing walls of ice four kilometers (two and a half miles) high. Each time, falling sea levels exposed the life-packed continental shelves to open air, disrupting marine ecosystems. Inevitably the glacial advance would wane, the walls of ice would retreat to the poles, and marine and terrestrial life would once again thrive in an interglacial period.

For a hundred million years, throughout the Carboniferous and most of the following geological period, the Permian, Earth's ice caps endured, occasionally sending glaciers down from the poles. Those ice caps finally melted away around 260 million years ago, when increased volcanic activity and decreased oceanic absorption of carbon rapidly pumped atmospheric CO_2 back to mid-Devonian levels. Abundant polar ice would not return to our planet until around 35 million years ago. Those polar ice sheets expanded just over two and a half million years ago, when the outpourings of undersea volcanoes formed the

Isthmus of Panama and sutured together North and South America, creating new oceanic and atmospheric circulation patterns that further lowered global temperatures. This occurred at the dawn of the Quaternary Period, a span of time that, at its tail end, would give rise to anatomically modern humans. Since the Quaternary's beginning, and even today, with Antarctica and Greenland still locked in ice, the Earth has technically been in an ice age. That this is a rather remarkable state of affairs has only very recently come to be appreciated. Polar ice caps, despite their presence for the entirety of human history, are surprisingly infrequent occurrences in the history of Earth. As far as geologists can discern, over the course of its 4.5-billion-year existence, ice caps have graced our planet's poles for only a sum total of about 600 million years—about an eighth of the Earth's life thus far.

In our present ice age, glacial walls of ice repeatedly pulsed from the Arctic to cover the sites of modern-day Toronto, New York City, and Chicago, as well as much of northern Pennsylvania. They carved out Hudson Bay and the Great Lakes, and at their edges spat out glacial moraines—chunks of broken land that became places such as Long Island and Cape Cod. The glaciers last retreated some 12,000 years ago, at the beginning of an interglacial epoch we call the Holocene. The rise of agriculture, cities, commerce, industry, science, and technology that we recognize as human civilization and chronicle as human history has all occurred within the abnormally mild and stable Holocene interglacial, the climatic equivalent of a twelve-thousand-year summer.

The signs of glacial advance and retreat can be tracked in sedimentary rocks and isotopic analysis of seawater, but some of the most high-fidelity evidence of climate oscillations comes from within the glaciers themselves, in bubbles of trapped, ancient air. Found in ice cores extracted from today's melting glaciers, each bubble is a snapshot of the atmosphere on a day in the distant past, when a minuscule puff of air was trapped in fresh-fallen snow that became part of the ice. The oldest bubbles are of impressive vintage—detailed analysis revealed they formed some 800,000 years ago. In aggregate, the gas in the bubbles tracked the changing composition of Earth's atmosphere for the majority of the past million years, long before modern humans ambled onto the planetary scene. The bubbles display a clear pattern that connects greenhouse-gas levels to glaciation: when glaciers were advancing, out of every million molecules of air trapped in their ice approximately 200 were CO_2. When glaciers were retreating, the amount of CO_2 in ice-locked air rose to about 300 parts per million (ppm).

Since the last glacial retreat at the beginning of the Holocene, circa 10,000 B.C., the average amount of CO_2 in the atmosphere had held steady at around 275 ppm. Glacial ice cores show that value beginning to climb at the turn of the nineteenth century, just as the world's population of humans reached one billion. Not coincidentally, this was when societies were embroiled in the Industrial Revolution, powering newly commercialized steam engines by burning ancient stores of coal. In all the time since, as populations surged and technology spread, CO_2 levels continued to sharply rise. By 1950, there were 2.5 billion people on the planet, and atmospheric CO_2 levels had surpassed 300 ppm—a level now considered the approximate threshold beyond which ice-age conditions cannot prevail. By 1980, the world held 4.5 billion people, and atmospheric CO_2 was 340 ppm.

By 2000, with the world host to more than 6 billion people, and with CO_2 at 370 ppm and rising some 2 ppm per year, a Nobel Prize–winning atmospheric chemist, Paul Crutzen, could no longer convince himself that he still lived in the Holocene. He had received his Nobel in 1995, based on his work clarifying how trace emissions of exotic gases—man-made refrigerants called chlorofluorocarbons—had in the latter half of the twentieth century eaten gaping holes in the planet's protective atmospheric layer of ozone. The ozone hole was only one small part of a much larger trend, Crutzen believed. Aided by engines and turbines, amplified by fossil carbon and petrochemical fertilizers, humans had commandeered huge swaths of the planet's flows of energy and nutrients, channeling them to new purposes and altering global geochemistry. The resulting growth was exponential, and—at least if confined to a single planet—unavoidably transitory. Geological history's darkest passages began to echo through his mind.

Evidence of the world's remaking was everywhere before his eyes. He could see it in the dissipating stratospheric ozone, the disappearing polar ice, and the thawing tundra permafrost. He glimpsed human hands in the shifting seasonal patterns of migratory animals and flowering plants, in the "once-in-a-century" storms, droughts, and heat waves that now came every few years, in the corals that lay bleached and dying in the too-warm waters of shallow seas, in the clear-cut forests, dammed rivers, and runoff-clogged streams. Crutzen felt compelled to coauthor an influential paper with the aquatic ecologist Eugene Stoermer arguing that our newfound planetary powers placed us in an entirely new geological period. In this "recent age of man," the "Anthropocene," human dominance altered the skies and seas, and even changed the very rocks that through the eons would endure as mute testament to our era.

For many millions of years to come, the Holocene-Anthropocene transition will be clearly visible to the naked eye wherever a proper rock face is exposed. In marine basins where white carbonate sediment once settled to form limestone and chalk, CO_2-saturated ocean water will have become more acidic, depositing instead dark carbonate-depleted clays and muds. If atmospheric CO_2 continues to climb unabated, carbon-rich black shales could dramatically re-enter the geological record, as rising temperatures and sea levels once again lead to widespread deepwater anoxia. We cannot say, however, whether whatever may later find the Anthropocene's new shales will, as we did, extract and burn the accompanying deposits of oil and gas to build a global technological civilization. Fossils abutting the Holocene-Anthropocene transition will record a planetary mass extinction event, the sixth in Earth's history, one in which species-rich ecosystems developed over tens of millions of years suddenly, irretrievably vanished and were replaced with agrarian homogeneity. Upper Holocene fossil beds will reward any future paleontologists with finds such as lithified coral reefs and carbonized amphibians; lower Anthropocene beds will more likely offer corn cobs, cow bones, petrified oil palms, and perhaps even some remains of their human masters, to whose needs they were bent and bred. Very rarely, a warped, weird band of rusted metal-laced rock may be found—the remnants of some major coastal city long sunken and buried beneath sediments from a great ancient river. If you hope to appear as a fossil in some fractured far-future cliff face, you could do worse than a burial on the Mississippi Delta of slowly sinking New Orleans.

When I met Mike Arthur in his office in late October of 2011, atmospheric CO_2 was 390 ppm—and still rising. According to UN estimates, planet Earth was only days away from a baby's birth, probably somewhere in southeast Asia, that would bring the world population to seven billion people. Similar estimates charted a course for population to hit ten billion late in the twenty-first century. If the global spread of technology and commerce continued, a very large fraction of the future population would seek to enjoy lifestyles similar to that of most present-day Americans. They would want coal-fired power plants to supply electricity at the flip of a switch, ubiquitous freeways and air travel, a car in every garage, a flat-screen television in every living room, cheap, disposable smartphones and computers, meat at every meal, and fresh produce flown to their dinner tables from halfway round the world. But barring the world's energy infrastructures somehow turning away from fossil fuels with unthinkable swiftness, those very behaviors would raise CO_2 concentrations beyond 500 ppm to reach 1,000 ppm and beyond—with predictably disastrous results. Average global temperatures could rise by 5 to 10 degrees Celsius, the poles would become wholly ice-free, and the rising seas would surge hundreds of kilometers inland around the coasts, to list only the most proximate effects. The Earth would then far more resemble the hothouse world it was during the reigns of the trilobites or the dinosaurs, rather than the cool and sedate planet that had previously nurtured human civilization. In that fevered future, simple survival of individuals and entire societies would become a constant struggle.

Whether all this would happen depended a good bit on when and how the Marcellus and other gas shales were exploited.

"As a scientist, I try to be objective, to base my conclusions on data and tradeoffs," Arthur said. "The data tell us that greenhouse gases from all the fossil fuels we're burning are greatly impacting the world. That is indisputable. It's also indisputable that burning natural gas produces 30 percent less CO_2 per unit of energy than oil, 40 percent less than coal. Coal-fired power plants produce more than half of the electricity in the U.S. People talk about converting to completely 'clean coal,' but there really is no such thing; mining coal is fundamentally deleterious to the environment. People talk about ethanol from corn or sugarcane. That's bullshit, a pig in a poke someone sold us. Natural gas is real, it's cheap, and it's the cleanest fossil fuel. If we replace a lot of our coal burning with natural gas, we're not just reducing CO_2; we're reducing emissions of mercury, nitrous oxide, sulfur dioxide, and particulates. We can make cars run on natural gas, too. It's not hard. That reduces emissions even more. So perhaps this is the lesser evil. But then I look at the sheer amount available, and I worry."

The Marcellus is singular only in its size. Similar black shales are found throughout the world, on every continent, each an echo and a portent of an Earth warmer and wetter than our own. A deeper, older deposit, the Utica, lies directly beneath the Marcellus in the northeastern United States. There are rich gas shales in Canada, Mexico, and Argentina. There are gas shales in Australia, China, and India. Gas shale is found across Europe—in Germany, in Poland, in the Czech Republic. Gas shale is in Africa, north and south. Gas shale seems, in fact, to be so abundant that it could almost unilaterally transform the fortunes of many developing countries, bringing economic prosperity and soaring levels of consumption and greenhouse emissions. In developed nations already hooked on ancient carbon, the shales could be a lifeline, found and seized just before the Anthropocene's fossil-fuel boom would otherwise reach its end.

Arthur looked down at his desk, strewn with unsteady piles of

paper, charts of geologic sections and maps of stratigraphic thickness—
the detritus of a scientist working to unwrap rockbound gifts from the
Earth. He briefly closed his eyes and raised his fingers to rub his forehead, as if to stave off the full-steam arrival of a freight-train headache.

Sooner or later, I think we will take most of the shale gas out of
the ground and burn it up," Arthur concluded. "Some people say it will
ease our transition into the other alternative energy solutions we need.
I worry it might stifle them instead. The view of most conservative politicians in this country today is, hey, why bother investing in solar and
wind and other renewables when we've got all this gas in the here and
now? 'Drill, baby, drill,' right? Well, let's say [the] most optimistic [gas
recovery] estimates are right, and let's say the U.S. for some reason used
the Marcellus as its only energy source. At present rates we would burn
through all of it in twenty years. Maybe it will last ten times as long, two
hundred years. Maybe longer. In the here and now that seems like a
good while, but remember, it took about two million years for the entire
Marcellus deposition to occur. Geologically, that's really fast, but it's
still too long for most everyone to comprehend. Humans are now influ-
encing the planet on these larger timescales, but we don't seem to be
very good at planning and accounting for that fact. We ignore the les-
sons of the past and the prospects of the future at our own peril."

Pennsylvania's rocks contain a record going just past the beginning of
the Cambrian Period, some 542 million years ago, which itself marks
the beginning of the Phanerozoic Eon, the half-billion-year stretch of
time, up to and including today, wherein geologists can find fossils of
complex organisms. "Phanerozoic" is a Greek appellation, and roughly
translates to "visible life." This was when organisms first began building shells and skeletons, which are more easily preserved in rock. The
emergence of hard body parts was only one part of a larger surge of
biodiversity famously called the "Cambrian explosion." Within a span

of only five or ten million years, organisms larger than a few centimeters in size became commonplace, and physiological innovations as fundamental as spinal cords, jaws, gills, and intestines all emerged for the first time. Nearly every extant animal on Earth today traces its architecture back to this baffling profusion of form. Extending our view a few tens of millions of years deeper into the past, we find the first evidence of creatures like worms and jellyfish, and features such as nerves, muscles, eyes, and radial and bilateral body symmetry. In nearly all Earth's long history before then, our planet was a world of prokaryotes—single-celled microbes lacking a nucleus.

For decades, scientists have struggled to learn what this vanished, largely alien "Precambrian" world was like, and why it so suddenly changed. Clues can be found in Precambrian rocks, and most of them suggest that, once again, some potent interaction between life and its environment had acted to irreversibly transform the world. The "smoking gun" in this ancient mystery is the very air you breathe—the oxygen in Earth's atmosphere, which lies at the heart of our planet's rapid, primeval transition.

Investigating the Precambrian is inevitably challenged by its great distance from us in time—the older any particular rock is, the greater the chance it has at some point in the past been melted or cooked, wiping out nearly all information it might otherwise contain about ancient events and environmental conditions. Three entire eons are contained within Precambrian time—four if you count the formative Chaotian, the eon in which our solar system assembled from its primordial cloud of gas and dust. After the Chaotian, the Hadean Eon began with the great Moon-forming impact some 4.53 billion years ago, and ended nearly 700 million years later. Of the Hadean we know virtually nothing. The Earth must have been quite hot as it cooled from formation, yet a handful of very rare Hadean rocks contain trace evidence of liquid water, suggesting that even then our planet may have had scattered surface seas. The boundary between the Hadean and its successor, the Archean Eon, is not well defined, being set by the smeared-out occurrence of the Late Heavy Bombardment between 4.1 and 3.8 billion years ago, when our solar system's giant planets seem to have hurled huge volumes of asteroids and comets down into the inner solar system. The great impacts that ended the Hadean and commenced the Archean Eon also began our planet's record of sedimentary rock. Any Hadean sedimentary rocks—and any Hadean life along with it—did not survive the pulverization of the crust and flash-boiling of the planet's water. It is in the Archean rocks that scientists find the earliest evidence of life, and the beginning of Earth's metamorphosis into the planet we know today.

Early Archean rocks contain hints of oceans, full-fledged plate tectonics, and the gradual growth of continents, as well as small amounts of organic carbon unquestionably generated by photosynthetic microbes. One thing they do not contain, however, is any significant trace of atmospheric oxygen. The Archean world was a desolate place, dreary beneath a sky clouded with smog-like organic hazes that built up in the anoxic conditions. It was also probably warm: the dominant life forms were likely one broad class of prokaryotes, called methanogens, that gained energy from reacting hydrogen and CO_2 together to produce methane, a greenhouse gas that can have even more heat-trapping potency than CO_2. The gloomy global reign of the methanogens and other anaerobic microbes seems to have lasted approximately a billion years. It could have lasted far longer if not for the sudden appearance of a new life form: the photosynthetic cyanobacteria.

First appearing in rocks from the latter half of the Archean, cyanobacteria were sea-green prokaryotes that, like Devonian plants or Holocene people, would go on to profoundly alter the world. In this case, the cyanobacteria decisively shaped the subsequent evolution of all life on Earth, and defined the final eon of the Precambrian—the Proterozoic, the two-billion-year stretch of "early life" prior to the Phanerozoic. Unlike the photosynthetic life of the planet's previous billion years, which used sunlight to gain chemical energy from hydrogen, sulfur, iron, and various organic molecules, cyanobacteria evolved a metabolic pathway to use sunlight to split water, a substance that was far more abundant and offered more chemical energy. The pathway seems to be a fluke of evolution—as far as scientists can tell, it only emerged this single time throughout Earth's long history. Its most obvious innovation was chlorophyll, a distinctively green class of light-absorbing molecules that more efficiently absorbed sunlight than the more ancient photosynthetic pigments, which were often pink or purple in color. After using chlorophyll to channel sunlight into water, cyanobacteria combined the harvested hydrogen with CO_2 to synthesize sugars, and jettisoned the leftover oxygen. Cyanobacteria also possessed the rare ability to draw chemically inert nitrogen gas out of the air to incorporate into the biochemical building blocks of DNA and proteins. Armed with the capacity to produce their own fertilizer, cyanobacteria could uniquely thrive wherever water, CO_2, and sunlight were present, and were poised to quite literally conquer the world. Late Archean and early Proterozoic rocks show that's exactly what they did, flourishing in vast open-ocean blooms and in concentrated communal mats blanketing the shallows and shorelines.

By 2.4 billion years ago, the Earth's new masters produced oxygen so prodigiously that they began to irreversibly transform the planet, as iron that had been dissolved in ocean water oxidized, solidified, and precipitated to the seafloor. It settled in thick layers of ferric sludge that were destined to someday become engine blocks, skyscraper beams, and battleship hulls. Early on, most of that oxygen was mopped up through reactions with organic carbon, volatile gases from volcanoes, and the rusting oceans. The oxidized material sank to the bottom of the seas, creating stratified and stagnant global oceanic conditions similar to those that would, much later, form black shales like the Marcellus. Experts endlessly debate what decisively shifted the planet's geochemistry away from sequestering all the surplus oxygen, but the result is indisputable: over the course of a few hundred million years, most of the ocean became saturated with the stuff, and ever after the gas flowed up into the atmosphere. In the upper atmosphere, the oxygen molecules clumped together to form a layer of ozone that absorbed large portions of the Sun's biologically harmful ultraviolet radiation, shielding life far below on the planet's surface.

Oxygen's rise was the world's first great pollution crisis, long before the invention of internal combustion engines and chlorofluorocarbon-spewing refrigerators. Despite the benefit of the Earth's new ozone layer, Earth's oxygenation was an unmitigated ecological disaster for the anaerobic biosphere that had developed and flourished during the Archean. To those creatures, oxygen's extreme chemical reactivity made it a terrible poison. Untold numbers of microbial species were annihilated as oxygen suffused the planet, and most of the surviving anaerobes retreated from the sunlight, initially into the anoxic muds found in the dark bottoms of deep seas and lakes, and much later into the low-oxygen digestive tracts of complex animals, including humans. They lurk in both kinds of shelters to this day. Oxygen also almost spelled the end for the cyanobacteria that produced it—the associated declines of anaerobic methanogens and heat-trapping atmospheric methane sent global temperatures plummeting and caused at least three Proterozoic ice ages, the first at 2.4 billion years ago, the second at 750 million years ago, and the third around 600 million years ago. Each was so prolonged and severe that glaciers reached the equator, repeatedly locking the oceans beneath a crust of ice that nearly eradicated all photosynthetic surface life. Joseph Kirschvink, a Caltech geologist who helped discover evidence of these extreme Proterozoic glaciations, called them "Snowball Earth" events after the likely appearance of our planet from space during each glacial episode. Near the equator or isolated volcanic hotspots, the ice might have been only a few meters thick, translucent enough to allow a twilight glow into the otherwise light-starved oceans where life hung on by the slimmest of margins. That the ice eventually thawed each time is clear, for otherwise we would not be here.

From these calamities sprang new opportunities: each Proterozoic ice age placed immense evolutionary pressures on the biosphere while also ratcheting up the level of energy-stoking atmospheric oxygen. A lucky few anaerobes, by dint of mutation and natural selection, adapted to tolerate the newly oxygenated atmosphere and ocean. Some of these new breeds of aerobic prokaryotes, in fact, took revenge on their conquerors by engulfing the cyanobacteria into their bodies as cellular slaves, making oxygenic photosynthesis their own. This process, called endosymbiosis, was what gave rise to the first eukaryotes, cells with centralized nuclei and specialized cellular structures. Modern plants are green because their cells contain chlorophyll-filled "chloroplasts"—structures that are scarcely distinguishable from cyanobacteria. The cells of modern plants and animals alike also contain enclosed structures called mitochondria, which are the cellular components that allow all eukaryotes to draw metabolic energy from oxygen—that is, to breathe. Chloroplasts and mitochondria each carry DNA independent of their host organism, confirming that both are captive descendants of prokaryotes incorporated into eukaryotic cells sometime in the latter half of the Proterozoic.

By the conclusion of the last great Proterozoic ice age some 600 million years ago, atmospheric oxygen was approaching the levels of today, and a new breed of fledgling eukaryotes waited in the wings to exploit its immense power to release chemical energy. For the first time, multicellular creatures could draw enough power from the air itself to support large, active bodies. The stage was set for the explosive diversification and growth of life, the rise of complex plants and animals, the colonization of land, and the eventual emergence of humans. We have now arrived where this pocket history began, at the oldest rocks of Pennsylvania, the root of the Cambrian, the great transition between Earth's three eons of simple life and its subsequent half billion years of burgeoning biological complexity.

Even armed with this cause-and-effect chronology, it can be difficult to understand why the Earth wasn't always the place we see before us, and just how its transition from an alien, hostile planet took place. Deep planetary time, in all its vastness, is the most fundamental thing to recognize. A thousand years of shifting climate can produce a forest where a desert used to be. A million years of tectonic activity can thrust up a mountain from wide-open prairie plain. A hundred million years of evolutionary trial and error can transform a prokaryote into a eukaryote, or a mouse into a man. A billion years is time enough to entirely restructure the ways of the world. To most people, the Phanerozoic and the Proterozoic certainly *sound* the same, and the difference between a million and a billion years is less a quantity of time and more a matter of three extra zeroes at the end of a numeric string. But simple thought experiments reveal the truth.

Consider that the entire 542-million-year span of the Phanerozoic is only about an eighth of our planet's total history. Imagine, as the writer Bill Bryson has, stepping into a time machine to venture back to the Phanerozoic dawn at a rate of 1 year per second. After ninety minutes, you would find yourself in the Bronze Age, around the time of the construction of Stonehenge, the domestication of the horse, and the founding of Abrahamic religions. A day later, you would be in the middle of the Stone Age, just as small bands of foraging humans began to migrate out of Africa. To reach the beginning of the Cambrian, the base of the Phanerozoic, would take you about 17 years. Now remember that almost a decade of earlier Precambrian time underlies each and every passing Phanerozoic year—departing from the far-distant Cambrian, your year-per-second time machine would take another 125 years to transport you to our planet's first moments.

Or try mapping the Earth's 4.5 billion years onto a calendar year. The Precambrian commences with the New Year's Day coalescence of Earth from the primordial nebula and persists until the Cambrian explosion in mid-November. Life gets going sometime in late February, but cyanobacteria only begin pumping oxygen into Earth's atmosphere in mid-June. The Marcellus shale forms a few days after Thanksgiving, and Pennsylvania's coal measures are all laid down in the first week of December. The following week, dinosaurs appear, but they succumb to extinction by Christmas Day. Anatomically modern humans show up late to the party, just after a quarter to midnight on New Year's Eve. One minute before midnight, the last glacial pulse ebbs back to the poles and the Holocene interglacial begins. Approximately one second before midnight, Earth enters the Anthropocene.

The author John McPhee has devised a more visceral visualization: Throw your arms out wide to represent the span of all Earthly time. Our planet forms at the tip of your left hand's longest finger, and the Cambrian begins at the wrist of your right arm. The rise of complex life lies in the palm of your right hand, and, if you choose, you can wipe out all of human history "in a single stroke with a medium-grained nail file."

Deep time is something that even geologists and their generalist peers, the earth and planetary scientists, can never fully grow accustomed to. The sight of a fossilized form, perhaps the outline of a trilobite, a leaf, or a saurian footfall can still send a shiver through their bones, or excavate a trembling hollow in the chest that breath cannot fill. They can measure celestial motions and list Earth's lithic annals, and they can map that arcane knowledge onto familiar scales, but the humblest do not pretend that minds summoned from and returned to dust in a century's span can truly comprehend the solemn eons in their passage. Instead, they must in a way learn to stand outside of time, to become momentarily eternal. Their world acquires dual, overlapping dimensions—one ephemeral and obvious, the other enduring and hidden in plain view. A planet becomes a vast machine, or an organism, pursuing some impenetrable purpose through its continental collisions and volcanic outpourings. A man becomes a protein-sheathed splash of ocean raised from rock to breathe the sky, an eater of sun whose atoms were forged on an anvil of stars. Beholding the long evolutionary succession of Earthly empires that have come and gone, capped by a sliver of human existence that seems so easily shaved away, they perceive the breathtaking speed with which our species has stormed the world. Humanity's ascent is a sudden explosion, kindled in some sapient spark of self-reflection, bursting forth from savannah and cave to blaze through the biosphere and scatter technological shrapnel across the planet, then the solar system, bound for parts unknown. From the giant leap of consciousness alongside some melting glacier, it proved only a small step to human footprints on the Moon. The modern era, luminous and fleeting, flashes like lightning above the dark, abyssal eons of the abiding Earth. Immersed in a culture unaware of its own transience, students of geologic time see all this and wonder whether the human race will somehow abide, too.

Immiscible emotions emerge from contemplating the dual realities of modern life and deep time—strange amalgams of apathy and anxiety that resist easy dismissal. Against the rich pageant of a planet and its past, the brief activity of a human life shrinks toward futility, even as human habits and behaviors, human choices in aggregate, so forcefully send much of Earth's complex biosphere sliding into oblivion. Yet as wrenching as the Anthropocene's changes may be, their permanence is as questionable as man's dominion: to be sure, once extinct, a species cannot be readily resurrected, but there are few reasons to believe that with time the planet's biodiversity could not recover, just as it has in the past. It can be seen as a blessing that modern civilization, in all its power, must struggle to even slightly perturb the robust microbial world that forms the basal roots of the Tree of Life. If it were otherwise, we could far more definitively disrupt the biosphere. Most of the more abiotic changes—alterations in geochemistry, atmospheric and oceanic circulation patterns, and so on—will eventually be reversed and erased by growing continental cratons, subducting oceanic crust, and erupting volcanoes. The fact that Earth's renewal will only unfold over millions of years may be no consolation to people today and tomorrow, no help to the countless species trampled beneath our civilization's tread, but that makes a recovery no less plausible. The grass will grow, the Sun will shine, and life on Earth will go on, with or without a wily band of tool-using primates. At least, that is, until the Sun, ever brightening through the eons as it fuses itself to death, brings an ultimate end to all things Earthly.

Whether all this makes the proximate probability of civilization's fall and the present destruction of the planet's biological wealth something to become upset about depends, therefore, on your sense of scale, and on where you think humans reside in what might charitably be called the "Big Picture," the assessment of things starting with the biosphere and extending out into the Milky Way. It is, in truth, only a minuscule fraction of nature's much larger tableau. Further up the cosmic scale, where even an entire galaxy is but one nebular mote out of hundreds of billions, or further down, to the quantum world of fundamental particles, the significance of Earth's spark of life, sentience, and technology grows indiscernible. But in all the uncertain sun-filled space between, it becomes just possible to see the promise of greater things, to envision our spark surpassing its billions of years of solitude upon a single planet circling a single star, flaring in ascension beyond planetary and stellar time to shine at the base of the endless, enduring galactic range.

CHAPTER 7

Out of Equilibrium

An earth scientist's tendency to see the Big Picture at the expense of smaller details helps explain something that happened to me one morning in the red-brick Deike Building that houses the Marcellus Center and Penn State's geosciences department. I was standing next to a bank of elevators in an otherwise empty corridor, waiting to meet someone. A short, bespectacled man in a button-down flannel shirt and khaki pants rounded a corner, glanced at me as he walked past, and entered a nearby bathroom. A minute passed, then the man reemerged and walked by me again, pausing to sip from a water fountain before making to return down the hall. When he was a few steps from vanishing around the corner, I called out to him, and when he turned around to look he didn't seem to immediately recognize me.

I was surprised, since I recognized him as Jim Kasting, a geosciences professor at Penn State specializing in the evolution of Earth's atmosphere and climate. After more than two hours of conversation in a noisy bar/restaurant called "Mad Mex" the previous evening, we had agreed to meet that morning to talk more about his work. I had spoken with him on the phone minutes earlier when I had arrived at Deike, but when he passed me twice in the hall I might as well have been one of the specimens of sedimentary rock in the glass cases that lined the wall.

"Oh," he said at last. "Hi, Lee. Didn't see you there. Let's go into my office."

Picture a NASA astronaut—not so much the stereotypical fighter-plane jock from the Space Race, but the post-Apollo variety, a strait-laced fitness buff with an advanced academic pedigree—and you probably have summoned a good approximation of Jim Kasting. Kasting is fifty-eight but looks years younger thanks to a strict regimen of swimming, running, and weight lifting. He is bookishly handsome, with a wide, magisterial forehead and the compact, sinewy build of a wrestler. He is equally at home discussing either the finer points of planetary carbon cycles or the benefits of rear-wheel drive on sports cars. Kasting speaks with clipped precision, and emotion rarely shades his voice. He never seems to be in any big hurry yet manages to be monumentally productive. His most astronautical quality, however, is something more subtle: a serenity that suggests awareness of one's inescapably small place in the world, an acceptance awakened by long hours spent contemplating the Earth from some lofty perch.

Kasting's resemblance to an astronaut was apt given his upbringing, which we had discussed over a cacophony of drunken Penn State coeds and a dinner of one-dollar tacos the previous night. He and his identical twin brother, Jerry, were born in Schenectady, New York, in the wee hours of January 2 in 1953. A younger sister, Sandy, arrived years later. His mother stayed at home raising her children, though she would later use her degrees in chemistry and mathematics to teach college courses. His father was a mechanical and electrical engineer, building jet engines as a subcontractor for General Electric. The family rarely stayed one place long, as GE moved them around the country to wherever the next contract was—first Schenectady, then Cincinnati, then Schenectady again—until 1963, when the family moved to Huntsville, Alabama, where they would stay for the next seven years. This contract was for something entirely new in the world, particularly for the Kasting brothers, who were in the midst of fifth grade: GE had sent their father to Alabama to work on third-stage engines for NASA's giant Saturn rockets.

In the 1960s Huntsville was a town dominated by the early promise of the Space Age. The rockets for America's first ballistic missiles, satellites, and astronauts had been developed at the nearby Redstone Arsenal, and most Huntsville families put their food on the table, directly or indirectly, with space-program funds. Out at a restaurant for supper, they might look over to see Wernher von Braun, the chief architect of the Apollo program, seated at an adjacent table dourly tearing into a steak. When they went home and watched the nightly news, von Braun would be there again on the television screen, speaking in Teutonic tones about the new high frontier. He headed NASA's Marshall Space Flight Center some twelve miles southwest of Huntsville. When every now and then Jim and Jerry would see a line of black limousines speeding through town, they knew another VIP federal motorcade was bound for Marshall and von Braun. America was going to the Moon, and the world seemed poised at the brink of a new revolutionary era. But the boys didn't truly appreciate the magnitude of their father's work until Huntsville began to regularly quake for minutes at a time. Bolted to static test stands down at Marshall, the Saturn rocket's great engines were being put through their paces, combusting huge reservoirs of liquid hydrogen and oxygen to produce millions of pounds of thrust per second. Each test firing began with a deep rumble that quivered the dogwoods and magnolias and clouded the sky with startled birds. The rumble rapidly crescendoed into a sustained roar that rolled beneath and through the town, cracking windowpanes and windshields and stirring a yearning in young Jim to, someday, work for NASA, if not as an astronaut, then maybe as a scientist. The roar of rockets signaled a future where humanity's fortunes would be found beyond Earth's cradle.

Kasting began working hard in math and science at school, and read all the science fiction he could get his hands on. One of his favorites was Isaac Asimov's Foundation series, books about the rise and fall of a galactic empire. Much of the story revolved around the empire's capitol planet of Trantor, a stand-in for what, at the time, passed as a plausible guess at Earth's not-too-distant future: a planet where land and sea, where nature itself, had been wholly smothered and subdued beneath the footprints of forty billion people and a glittering techno-utopia of skyscrapers, superhighways, and domed farms and habitats. "I liked books with big ideas, ones that dealt with the future of humanity, or how to run a society," Kasting told me. "*Foundation* had a cool one: 'psychohistory,' the idea that if you have enough people they will behave just like atoms or molecules, individually unpredictable but foreseeable in aggregate, so a civilization's behavior becomes like that of an ideal gas, something controlled through statistical mechanics. I don't know if that's true—people are pretty complicated—but it made me think more about what can be predicted."

Late one evening when the boys were in middle school, Kasting's father arrived home from work with a tripod-mounted 2.5-inch refractor telescope, suitable for viewing all that the new rockets were bringing into reach. On dark, clear nights, they could see Saturn's rings, the ruddy disk of Mars, and the plains and craters of the Moon where men soon would walk. Through the viewfinder, the jagged, magnified lunar surface looked close enough to touch, like some monochrome impasto landscape hung on a museum wall. Jim's interests exceeded the limits of the solar system a few years later when he upgraded to a more powerful 4.25-inch reflector, and he began searching the sky for nearby planetary nebulae and neighboring galaxies. Sometimes he would wonder how the Earth or another inhabited world would appear, viewed from so very far away, if only there was a telescope big enough to look.

After high school, Jim plotted a trajectory he hoped would intersect with NASA's orbit: undergrad at Harvard, then a PhD in atmospheric science at the University of Michigan, and finally a series of postdoctoral positions. In 1981, he achieved his dream, securing a research fellowship at NASA's Ames Research Center in Mountain View, California.

Not long after Jim's NASA debut, his father paid him a visit out in California. By then, Jim had met and married his wife, Sharon, and their first child, a son, Jeff, had just been born. Kasting's father listened attentively, smiling and nodding as Jim showed off his burgeoning efforts to model the early atmospheric evolution of Venus, Earth, and Mars—working on the problems full-time with NASA's winds at his back, he was making progress fast, and getting further than anyone had before. Perhaps skeptical that Jim could raise a family by predicting a planet's far-distant past and future, or maybe just habituated to always push for greatness, when Jim had finished explaining Kasting the elder promptly asked his son when he planned to get a real job. In fact, Kasting's work had already begun to revolutionize planetary science and had placed him on the NASA fast track. In 1983, when his fellowship expired, he was immediately hired as a research scientist at Ames, where he would remain until his 1988 emigration to Penn State. With NASA money as their nest egg, Jim and Sharon would have two more sons, Patrick and Mark.

Kasting's Penn State office was adorned only with a blue-and-white Oriental rug and a few yellowing astronomy-themed posters that broke the spartan regiments of books, papers, and reports. One side of the room was occupied by three large filing cabinets, collectively filled with a good half ton's worth of astrobiology's primary literature. The other side was taken up by bookshelves mounted on the cinderblock walls. The shelves brimmed with well-thumbed, dog-eared volumes with such titles as *Biogeochemistry of Global Change, The Chemical Evolution of the Atmosphere and Oceans*, and *Fundamentals of Atmospheric Radiation*. An adjacent whiteboard was filled top to bottom with scribbled shorthand references to stellar flux, atmospheric partial pressures, and surface temperature, as well as three frenzied, overlapping strata of differential equations, each distinguished by its own shade of erasable marker.

The books and equations revealed Kasting's true interests, which go far beyond our own small planet and its history, harking back to his musings at a backyard telescope. He is widely considered the world's foremost authority on planetary habitability—how a life-friendly planet can emerge and evolve over geologic time. Like the Earth itself, he has spent most of his time within the Precambrian's murky frontiers. Among other things, he has had a hand in calculating how much longer photosynthesis can support complex life on Earth (about a billion years), the minimum size an impacting asteroid must be to vaporize Earth's oceans (one 270 miles wide would do the trick), and whether by burning all available fossil fuels humans could force the Earth into a Venus-style runaway greenhouse (the jury is technically still out, but Kasting believes the answer is, thankfully, "no").

Over dinner the night before, I had suggested that we hike through some of the surrounding Pennsylvania wilderness, so that Kasting could use examples from the landscape to illustrate his Big Picture view of Earth as a system, of habitability as a process unfolding over geologic time. "If you take me out in the field, I'm pretty useless," he initially demurred. "I've actually had no formal education in geology. I probably couldn't tell you if a rock was a carbonate or a silicate. I'd be lucky to know a glacial till from a landfill." After finishing a margarita, he had changed his mind, and offered to take me to Black Moshannon State Park, five square miles of forest and wetland located a twenty-minute drive northwest from Penn State's campus. "I still won't be very useful," Kasting said, "but it will be a nice walk."

• • •

Behind each modern announcement that scientists have found yet another possibly habitable world is a well-worn process that, simplified, unfolds as follows: Astronomers first measure the newly discovered planet's mass, and, if possible, its radius, generating an estimate of the planet's density and its likelihood of being rocky like the Earth. They also determine the rocky planet's orbital distance from its star, as well as the intensity and color of the star's light. Armed with this scant data, the entirety of which you could jot in ballpoint pen on the palm of one hand, they then interpret it through numerical modeling. In particular, they consult one of Kasting's most-cited papers, "Habitable Zones around Main Sequence Stars," published in the journal *Icarus* in 1993. In that paper, Kasting and two colleagues, Dan Whitmire and Ray Reynolds, used a climate model developed by Kasting to determine which orbits around stars are most likely to allow rocky planets to harbor liquid water upon their surfaces. Inward of the habitable zone, a planet's surface would be so scorched that any water would flash to steam, suffusing the atmosphere and gradually escaping into space, similar to what occurred on Venus; outward of the zone, a planet's surface water would freeze, similar to what we see on Mars. If a newfound rocky planet proves to be within Kasting's habitable zone, shortly thereafter its discoverers contact their funding institution's press office, and soon their names appear on the nightly news and in the *New York Times*. Kasting coauthored a paper in January 2013 gently revising his twenty-year-old calculations, but the tweaks did not greatly alter his earlier work's core conclusions.

Using a literal handful of data points to estimate the habitability of a faraway planet is a practice fraught with uncertainty, where major assumptions and leaps of faith become inevitably routine. That it is possible at all is only because, as far as we can tell, the laws of nature are everywhere the same throughout our observable universe, whether in the solar system or around some far-distant alien star. Anywhere in the universe where starlight falls upon a planet, it pumps radiant energy into the system of that world. How much energy filters in depends on the planet's atmosphere, and upon the starlight's wavelength, or color. For those canonical 1993 calculations, Kasting and his colleagues gave their virtual planets atmospheric compositions thought to be the most typical outcome of terrestrial planet formation: lots of inert nitrogen, accompanied by substantial fractions of CO_2 and water vapor. Evidence suggests this was the bulk atmosphere of the early Hadean Earth, but for distant rocky exoplanets with atmospheres that have yet to be measured, any particular mix can presently be seen as only a hopeful guess.

After a particular atmospheric cocktail is chosen, the core of Kasting's numerical approach kicks in, most of which he developed during his seven years at NASA. During all that time, he devoted himself to perfecting his models, hand-coding each important way that starlight interacts with an atmosphere. In the real world, and in Kasting's models, a photon of a certain wavelength might simply bounce off the top of the atmosphere, while a photon of another wavelength might instead pass without incident all the way down to the planetary surface. Inside the atmosphere, real or virtual, a photon might be reflected by a cloud, or by bright ice on the ground. It might be absorbed by a greenhouse gas, or by the dark water of a sea. When a photon is particularly energetic—ultraviolet or higher on the electromagnetic spectrum—it might even create entirely new substances in the air and on the ground by knocking into molecules and splitting them apart—a process called "photolysis." The photolytic products could then have their own secondary effects on the absorption and reflection of starlight, all of which must be taken into account. Over the years, Kasting accumulated all the necessary data he could find, building up a vast library of radiation-absorption tables, photochemical reaction rates, atmospheric lifetimes of different gases, and the global pace at which certain gases arc emitted from volcanoes or absorbed by rocks. Collectively, all these various interactions and inputs have an enormous effect upon a planet's atmospheric composition and average surface temperature—its climate.

If you naively calculated the average temperature of the modern Earth's surface based only on the amount of sunlight it receives and its average reflectivity, or albedo, you'd obtain a value of –18 degrees Celsius, well below the freezing point of water. If you calculated it using one of Kasting's climate models, you'd get a result of 15 degrees Celsius, which is, of course, what the Earth's average surface temperature actually is. The discrepancy is mostly due to warming from several different greenhouse gases, each of which Kasting must account for in painstaking detail.

Water vapor, for instance, must be treated very carefully, as it is actually a much more potent greenhouse gas than CO_2, efficiently absorbing a much broader swath of the thermal-infrared portion of the spectrum. Further, its effect on climate is qualitatively different: unlike CO_2, which stays gaseous at typical Earth temperatures, water vapor is intimately affected by Earth's temperature changes. Low temperatures will cause it to condense into clouds and fall out of the sky as rain, snow, and hail, which removes its greenhouse effect and drives temperatures even lower. Conversely, high temperatures increase the evaporation rate of surface water, pumping more water vapor into the air to raise temperatures even further. Water vapor thus acts in a positive feedback loop to amplify other climate changes, such as the steady heating forced by rising levels of atmospheric CO_2. If CO_2 is the fulcrum about which Earth's climate change pivots, water vapor is the lever.

The key output of one of Kasting's climate models is something called a temperature-pressure profile—scientific jargon for how starlight shining on any given atmosphere will influence not only its warmth, but also its vertical structure. Earth's atmosphere, for instance, reflects a quarter of the incoming sunlight and absorbs another quarter through greenhouse gases, allowing approximately half of the sunlight that strikes it to filter down to the surface. This means that, on average, Earth's atmosphere is colder than its surface, and is warmed from the bottom up by convection, like a pot of water being heated on a stovetop. Most of the surface heating and convection occurs around the equator, where, as a cursory examination of any globe will show, there is more surface area to absorb the sunlight that beats down from almost directly overhead. Convective cells of moist air undulate from the warm surface, cooling as they rise and expand, eventually growing cold enough to dump their moisture as condensed water vapor—that is, as clouds and rain. Atmospheric convection helps to explain why the tropics are hotter than the poles, why air around high mountaintops, though fractionally closer to the Sun's radiance, tends to be thinner, colder, and drier than the air at sea-level plains, and why thunderstorms typically occur on torrid afternoons and early evenings, hours after the Sun's zenith.

Earth's temperature-pressure profile creates a feature in the atmosphere called the tropopause, a dividing line which runs above the warm, weather-filled troposphere and below the colder, thinner stratosphere. Since water vapor condenses when exposed to cold temperatures, it is effectively trapped beneath the tropopause by the colder overlying atmospheric layers. Just how important this "cold trap" effect is for Earth's prolonged possession of water became apparent in the 1980s through a series of studies by Kasting, his colleague James Pollack, and a handful of their peers at NASA Ames. They were interested in understanding why Venus, our planet's near twin, had developed such a dramatically different climate than Earth, despite evidence that in its very early history our sister planet had been hospitably tepid and wet, rather like our own world now.

"To someone like me, the single most interesting thing about Venus is what it says about the inner boundary of the habitable zone," Kasting explained as we chatted in his office. "It sets a reasonable empirical limit on what you can expect for other planets outside the solar system—you don't need to do much modeling to guess that something getting Venus's amount of starlight probably won't be habitable. So if you want to know what happens when an otherwise Earth-like planet forms too close to its star, or what can happen to a habitable planet as its star gets brighter over time, Venus can tell you a lot."

Building on previous work performed by several other planetary scientists, most notably Caltech's Andrew Ingersoll, Kasting modeled how the Earth's atmospheric structure—Earth's temperature-pressure profile—would react to increased intensity of sunlight, as would occur if the Earth's orbit were moved in toward the Sun to a more Venusian orbit, or when the Sun slowly increases its luminosity over geological time. He found that with a relatively modest 10 percent increase in the starlight's intensity, equivalent to moving the orbit of our planet to 0.95 AU, 5 percent closer to the Sun, the additional warming would saturate the troposphere with water vapor, pushing the tropopause up to altitudes of 90 miles or more.

As Kasting watched the tropopause soar in his numerical model, he knew he was witnessing what would lead to the end of that virtual world, and someday, our own: much of the water vapor that lofted to such heights would rise above the protective ozone layer, where it would be photolyzed by ultraviolet light from the Sun. A small percentage of the liberated atomic hydrogen would escape entirely into outer space, taking with it any potential to ever bond again with Earthbound oxygen to create water. Within a few hundred million years, enough hydrogen would be lost to space in this manner that Earth's oceans would essentially boil away, leaving the planet lifeless and dry as a bone, with not a drop of water left upon its surface or in the air. In a billion years, long before it swells into a red giant and threatens to physically engulf our world, the Sun will have brightened by that crucial 10 percent, and the Earth will begin to rapidly lose its water and its life. This "moist stratosphere" mechanism is now thought to be how Venus began losing its oceans early in our solar system's history, and its threshold of 0.95 AU for our own planet conservatively approximates the inner edge of Kasting's habitable zone from his canonical 1993 paper.

As Venus lost its oceans, the rising temperatures baked CO_2 out of the planet's crust, and the gas began to fill the atmosphere. As a result, Venus's atmosphere is now some 90 times denser than Earth's and almost pure CO_2, creating a greenhouse effect so potent that the planet's surface temperature is hot enough to melt lead. In a second series of studies, Kasting and his coworkers modulated the CO_2 content of Earth's atmosphere to examine whether increased CO_2, rather than increased sunlight, could on a much faster timescale independently lead to the loss of oceans through a moistened stratosphere.

To his surprise, Kasting found that even as rising CO_2 levels sent temperatures skyrocketing, the vast amounts of water vapor released acted like the lid on a pressure cooker, pressurizing the lower atmosphere to such an extent that the oceans never boiled, keeping Earth's stratosphere relatively dry. For the stratosphere to become saturated with moisture, for the oceans to vaporize and escape into space, the numerical models indicated that Earth's atmospheric CO_2 would have to reach more than twenty-five times its present concentration—more than could be released by burning the entirety of our planet's known "conventional" fossil fuel reserves of oil and coal, but just maybe within reach if all the planet's unconventional sources, like the Marcellus's shale gas, were burned as well. While humanity could readily give the planet a fever that could shrivel societies and severely diminish existing biodiversity, Kasting's calculations suggested it would be very much harder—though not definitively impossible—for humans to create a moist stratosphere. By his reckoning, forcing the planet to give up its ocean to space by burning fossil fuels appears to be just beyond the reach of present-day civilization.

There are, however, significant uncertainties in Kasting's considerations, such that science cannot yet entirely dismiss the possibility of a man-made moist stratosphere leading to a premature runaway greenhouse on Earth. Other greenhouse gases besides CO_2 and water vapor play a role in Earth's climate, and could potentially have significant future effects that are unaccounted for in Kasting's models. And no one presently knows the exact amount of fossil fuels locked away within the Earth, or how much of that guesstimated total could be effectively extracted and burned based on future market conditions and potential technological development. Most fundamentally, no one fully understands how wide variations in temperature and pressure can subtly affect water vapor's absorption of thermal-infrared radiation. Nowhere is this haziness more evident than in considering the problem of clouds.

To the average person, clouds are simple things, pieces of cottony fluff in blue skies or ominous gray sheets portending dismal weather. To a climate modeler like Kasting, clouds are the most mercurial and beguiling form of water vapor, fickle creatures almost alive in their fiendish complexity. Depending on a cloud layer's extent, altitude, and composition, it may either warm or cool a planet. A blanket of dense, low clouds can reflect a good portion of sunlight into space, potentially reducing temperatures. But throw a layer of thin clouds high above the low, dense ones, and much of that cooling effect will be undone, as the translucent upper layer now allows sunlight to stream down but traps the heat that subsequently tries to escape. What everyone agrees on is that as a planet like Earth warms, more water vapor steams into the air to form more clouds. But there is no consensus on where exactly those clouds would form and linger in the atmosphere, or the limits of their feedback effects. Both global-warming deniers and publicity-hungry planet hunters have found refuge in the resulting nebulosity: water-vapor clouds could, in theory, save an otherwise habitable planet from runaway global warming, whether induced by an overabundance of greenhouse gases or by the too-bright light of a nearby star. Farther out away from a star, where temperatures drop low enough for CO_2 to condense into ice, an insulating blanket of dry-ice clouds could in some circumstances warm a planet enough to preserve liquid water at its surface. In 1993, Kasting conservatively estimated the habitable zone's outer edge to lie slightly beyond the orbit of Mars at 1.65 AU, but it could in fact extend out much farther, depending in large part on the uncertainties associated with CO_2 clouds.

There are two divergent strategies for numerically approximating clouds. One is to model them as accurately as possible in extremely detailed three-dimensional simulations. This approach requires reams of data from Earth-observing satellites as well as state-of-the-art super-computers, and risks losing the distinction between cause and effect in a flurry of variables and feedbacks. The other strategy is to model clouds much more simply in fewer dimensions, which carries the risk of overlooking vital behaviors that only emerge through complex inter-actions beyond the model's boundaries. Kasting prefers simplicity. His models are one-dimensional, approximating the entirety of a planet's atmosphere with a single linear sounding, something like measuring the average temperature and salinity of an ocean by sampling seawater through a very long seabed-to-surface soda straw.

"Clouds are pretty arbitrary in 1-D—you can get any effect you like in a 1-D model by playing with how you represent them. The ideal sce-nario for a 1-D model is a cloudless sky, which is obviously a huge weak-ness," Kasting acknowledged when we discussed his models. "I try to get around it by basically painting the clouds on the ground, approximating their effect by tuning the surface albedo until it reproduces the average temperature of whatever planet I'm trying to look at—Earth, for example, or Mars. Some people don't like that, and exactly what my method actu-ally means in terms of real clouds is complicated, but I think of it as min-imizing any cloud feedbacks that may occur as a planet's temperature changes. To do any better than that, you have to go to 3-D, which is a very big step, and even there, clouds remain the biggest uncertainty—the 3-D guys don't know how to do them, either."

Owing to its simplicity, a 1-D model is also much faster than any 3-D counterpart. A state-of-the-art 3-D climate model might take a week on a very expensive dedicated computing cluster to arrive at the conclusion that doubling Earth's present atmospheric CO_2 levels would raise the average temperature somewhere between 2 and 5 de-grees Celsius. Kasting's 1-D climate model calculates the results of a CO_2 doubling in less than a minute on a run-of-the-mill desktop com-puter, and arrives at an answer of 2.5 degrees. "With a 1-D model, I'm limited by how fast I can think, not how fast my computer can,"

Kasting said. "So over the course of a week, while a 3-D model may be processing a single iteration, I can pretty well explore the entire parameter space. That's what this is about—exploring the limits of what appears possible, and challenging others to build on that numerically or to look deeper empirically."

The more we talked, the clearer it became that Kasting had grown increasingly jaded in recent years toward press releases claiming progress toward finding exoplanetary twins of Earth. Much of the early furor had centered around the planetary system of the red dwarf star Gliese 581, some 20 light-years from Earth. First came Gliese 581c, a super-Earth skirting the inner edge of Kasting's habitable zone that, for a few months in 2007, was thought to perhaps be clement. Simple calculations by Kasting and others, however, revealed that regardless of atmospheric composition, the planet is bathed in 30 percent more starlight than Venus. Attention then shifted to its farther-out companion, the super-Earth Gliese 581d, which brushes the habitable zone's outer edge. Its habitability could not be as easily ruled out as c's, but Kasting hastened to point out that the world receives 10 percent less starlight than Mars. Additionally, c and d are so massive, each more than five times the bulk of our own world, that they could well be gas-shrouded shrunken Neptunes rather than rocky super-size Earths. Then in 2010 came the announcement of Gliese 581g, Zarmina's World, orbiting in the middle of the habitable zone, and, with an estimated mass a bit more than three times that of Earth, almost certainly terrestrial. That one had excited Kasting, at least until other astronomers began questioning the planet's existence.

Some months before our meeting, a European team had announced the discovery of another potentially habitable super-Earth, HD 85512b. Kasting thought "potentially" was an overly generous descriptor—the planet roasts in only slightly less starlight than Venus. "They wrote that a ton of cloud coverage could reflect all that light and make things okay," he recalled, referring to the European team's discovery paper. "But clouds sure didn't save Venus, did they?"

News of more potentially habitable planets had by then become a fairly regular occurrence, with each world's fortunes cresting and subsiding on the shifting tides of public interest and scientific opinion. Each discovery followed a similar cycle, first announced in academic journals, the primary producers upon which much else feeds. Pure empirical measurements of masses, orbits, and stellar fluxes then filtered down to the hazy murk of news reports, which processed them into rampant speculation. An infectious cocktail of certain fact and wild conjecture about each promising world then fanned out to mutate within the darkness and confusion at the base of so much human discourse. Before long, bizarre blog and forum postings would appear, wondering when NASA would send a probe, or, better yet, colonists, and whether when we arrived we'd find the builders of the Egyptian pyramids, or perhaps the home planet of the cattle-mutilating, human-abducting Grays, or maybe even Jesus Christ, on yet another pit stop midway through his universal tour of salvation. Again and again, the handful of known facts for each planet became buried beneath the familiar fictions that so many people construct for themselves.

Seeing the same pattern play out for each successive world, Kasting and his peers sometimes felt like soothsayers, obediently displaying tea leaves, yarrow stalks, chicken entrails, and other crude omens to an audience eager to imbue them with subjective meaning. One researcher once told me, with pained exasperation, that provided with only a planet's mass, radius, and orbit around any given star, the best way to determine the world's actual surface temperature would be to grab a newspaper and consult the horoscopes.

Kasting was less extreme, but equally dismissive. "None of these announcements of planets in or near the habitable zone should be big news by themselves," he told me with an edge of frustration. "They are in a way meaningless, because we presently can't follow up on the initial act of discovery. The big news will only come when we are able to actually look at one of these planets to discern whether or not it is actually habitable and see if there is evidence of life, right? And if we do that—excuse me, *when* we do it—the real revolution will begin."

To that end, for most of the past two decades Kasting had devoted his time and effort to two intertwined tasks: how to distinguish whether any terrestrial planet is a living world or a lifeless rock based on only a faint smear of starlight reflected off its atmosphere, and how to design a space telescope capable of making those observations. He had worked tirelessly, serving on a multitude of planning committees, panels, and task forces for NASA, the NSF, and the National Academy of Sciences. He helped churn out a mountain of reports defining the observational criteria to which armies of engineers and mission planners would eventually aspire, and it is no exaggeration to say that for a time nearly every definitive paper on the subject bore his name as a coauthor. The telescopes Kasting wished to build were called Terrestrial Planet Finders—TPFs for short.

In the years leading up to the turn of the millennium, as the pace of exoplanet discovery quickened, American federal coffers, bursting with surplus, had generously funded all manner of space science. The quest for exoplanetary life, like the nation itself, had seemed set on an unstoppable upward trajectory; telescopes to gather evidence for or against other nearby living worlds would be in hand, Kasting and his peers had told themselves, within perhaps a decade. Instead, a series of catastrophes had soured the nation's fortunes and slowed meaningful progress to a virtual standstill. The terrorist attacks of 9/11, the ensuing ruinous wars and unbalanced federal budgets, the collapse of the housing-securities bubble, and the onset of the Great Recession all could be said to have played a role, but much of the blame for TPF's failure was due to territorial infighting between competing communities of astronomers scrabbling for dwindling federal funding.

"It's one of the few things I get upset about, because I used to hope that something like the TPFs would happen in my career," Kasting had confessed when he was halfway through his margarita the night before. "I no longer hope that. I now just hope it happens while I'm still alive, because I want to know the answer. But my personal time is running out, and it seems to be receding further and further into the future. There's a good chance a TPF-style mission won't happen until after I'm dead."

When he talked about looking for life-bearing exoplanets, Kasting's words sometimes took on martial tones. He would "fall on his sword" to keep NASA missions like Kepler operating as long as necessary to find Earth-size planets in habitable zones, and would "fight to the end" for bigger, better space telescopes to search for signs of life. The money required for a TPF, he pointed out with rancor, was large on astronomy's scales but trivial by national and international standards: five or ten billion dollars for the chance to find out if humans were not alone in the universe, equivalent to a few weeks' worth of war in the Middle East, less than a year of Americans' expenditures on pets. The astronomers were being kicked around by NASA, and NASA was being kicked around by a monumentally dysfunctional Congress. Not that astronomers were blameless: Kasting took a dim view of senior space scientists who still viewed the exoplanet boom with disdain. Discussing them, Kasting's once-fiery words suddenly were rimed with frost: "They are mostly old cosmologists, and ten years from now, a lot of them will have died. The young kids all surging into exoplanets now should eventually dominate the decision making. Statistically speaking, the opposition will be buried just by numbers."

For Kasting, if the search for habitable exoplanets was not something to die for, it certainly was worth the remainder of his days. And in that calculation, any difference became blurred. He didn't consciously dwell on his own mortality as he mechanically swam, ran, and lifted weights in his morning workout sessions, but in the back of his mind each stroke, stride, and bench press became an extension of life, a flint-spark struck against the onrushing night, propelling him incrementally further forward in time toward the elusive light of other living worlds. It wasn't selfishness that drove him, but fear—fear that when faced with the possible discovery of potential signs of life on an alien world, planet hunters would botch the call.

"I hate to say it, but most astronomers I've talked with have shown no evidence that they really know anything at all about planets," he had told me the previous evening as we finished our dinner. "If I'm still around when we get a potential hit, I can help determine if it's real, and if I'm not, hopefully my ideas will be." Kasting had hedged his bet on personal longevity by condensing and dumping his acquired knowledge into an instruction manual that would certainly outlive him: *How to Find a Habitable Planet*, which was published in 2010 by Princeton University Press.

Swigging melted ice from his now-empty margarita glass, Kasting excused himself and said it was time for him to get home: it was almost 11:00 p.m., but he planned to hole up in his study to prepare a lesson for an undergraduate class the following day, as well as a presentation for an upcoming meeting of NASA's Exoplanet Exploration Program Analysis Group, a top-level planning committee he chaired and viewed as perhaps his last chance to steer the agency's course toward his hoped-for TPF space telescopes.

Four months later, Kasting would step down from his chairmanship, driven out by critics who said any mission like TPF had slipped too far into the future to be worth seriously considering.

Proposals to look for chemical signs of life on other planets—"biosignatures"—first emerged in the summer of 1965, via two separate papers, both published in the journal *Nature*, a month apart. Both papers primarily concerned the search for life on Mars. The first was authored by Joshua Lederberg, the Nobel laureate chemist who had mused about the prevalence of extraterrestrial intelligence at Frank

Drake's Green Bank meeting four years earlier. In his paper, Lederberg laid out several guiding principles, among them the idea that life could be detected by its indirect thermodynamic effects upon a planet's environment. Any conceivable organism must metabolize to survive—that is, must draw energy from and eject waste into its environment to grow, reproduce, and maintain orderly structure. Life on Earth, and presumably all life based on chemicals, drives its metabolism using chemical-energy gradients that chemists call "redox reactions," in which electrons are transferred between substances. (If a substance gains electrons, it is counterintuitively said to be "reduced." If a substance loses electrons, a chemist would say it is "oxidized," even if no oxygen was involved in the reaction, because oxygen is one of the most voracious electron acceptors known. Confusing vagaries of nomenclature such as these are in large part why many science journalists avoid writing about chemistry.) Lederberg noted that metabolic processes, regardless of their biochemistry, should create extreme thermodynamic disequilibrium upon a planet. These global-scale chemical imbalances would be made by organisms locking away energy and vital molecules in biomass and ejecting degraded waste products. He wrote that searchers might generally look for a biosignature of "chemically unstable [molecules] which should reach equilibrium with coexistent oxidant," a thermodynamic miracle akin to finding an unblemished, flame-licked log at the heart of a roaring bonfire.

The second paper, by the British scientist James Lovelock, honed Lederberg's broad assertions into a far sharper criterion for life: a planet's atmosphere, Lovelock suggested, would be the best target to examine for signs of thermodynamic disequilibrium. In particular, the search should look for "the presence of compounds in the planet's atmosphere which are incompatible on a long-term basis." Lovelock cited the Earth's atmosphere as an example, since oxygen and methane both exist there in chemically implausible concentrations. If left alone in a sealed vessel at room temperature and pressure, oxygen will react with methane to form carbon dioxide and water. Yet in the Earth's atmosphere, which is composed of just over 20 percent oxygen by volume, methane somehow persists at just under 2 parts per million—the two gases are out of equilibrium by nearly 30 orders of magnitude. The sole explanation for this lingering thermodynamic imbalance is that methane is being constantly replenished.

Almost all of Earth's methane comes from our planet's ancient Archean refugees, the anaerobic methanogens, though a very small fraction is also produced abiotically by hydrothermal vents on the ocean floor. Even without the methane, Earth's abundant oxygen is by itself far out of equilibrium and exceedingly peculiar, as oxygen prefers to bond with rocks and minerals rather than linger in the air. It clearly must be replenished, too. Our world's oxygen, of course, primarily comes from photosynthetic bacteria and plants, though, as with methane, small amounts can be produced abiotically—in oxygen's case, when ultraviolet starlight photolyzes water vapor. Since both oxygen and methane have possible abiotic production routes, the existence of one without the other cannot necessarily be taken as a certain, foolproof biosignature. But when they appear together, their presence constitutes the most potent evidence astrobiologists can recognize for life beyond the solar system, short of a SETI-style radio transmission or a flying saucer landing on the White House lawn.

"It's simply very hard to abiotically build up large concentrations of both methane and oxygen in a planet's atmosphere," Kasting said to me later, as we began our drive to Black Moshannon. "Finding evidence of both in the atmosphere of a planet like ours—a rocky planet with surface water and enough mass to hold on to its internal heat and drive something like plate tectonics—would be, to me at least, an ironclad detection of life. People might say that's like looking for your lost keys only under street lamps, where the light is brightest, but I don't think that's exactly true. The truth is that the public never wants to hear some scientist rule something out entirely, so it's politically correct to say, 'Oh, these alien biospheres could of course have very different chemical signatures than anything we know here on Earth!' I think that's totally wrong. Maybe someday I'll eat my words, but I believe the only biosignatures we can reasonably look for are the ones that we can presently model and constrain—the ones we either see here on the present Earth or know the Earth had in its past. If life starts up on any planet that shares a few key characteristics with our own, I think you'll get a biosphere we can spectroscopically recognize. A metabolism will work the same even if the cells are very different than ours and don't depend on DNA and RNA molecules. Here or there, putting CO_2 and hydrogen together to drive a metabolism will produce methane. Here or there, taking hydrogen from water and venting out the oxygen is a winning metabolic strategy if life can figure out how to do it. Chemistry and thermodynamics are the same everywhere."

As elegant as Lovelock's criterion appears on paper, its great shortcoming is that the spectral signatures of oxygen and methane manifest at very different wavelengths. Oxygen absorbs starlight most efficiently in the near-infrared, creating prominent spectral "absorption bands" just outside the part of the spectrum that our eyes can see. Methane, being a very potent greenhouse gas, most efficiently absorbs at the longer wavelengths of the thermal-infrared. In astronomy, working at longer wavelengths translates to using bigger light-gathering areas—this is why radio telescopes are so much larger than optical ones. It's also why detecting both oxygen and methane in an exoplanet's atmosphere may well require the coordinated efforts of at least two space telescopes. One would be smaller and simpler, to observe oxygen in the visible and near-infrared, while the other would be larger and more complex, to observe methane in the thermal-infrared. Working together, the telescopes could also measure other gases in a planet's atmosphere, notably water vapor and CO_2, which would help constrain a world's habitability and climate—extreme amounts of either gas would suggest a world too hot to support liquid water and life, while more moderate amounts would indicate the presence of surface water and more hospitable surface temperatures.

"This won't be an instantaneous thing, because planning and

building multiple large space telescopes takes a lot of time," Kasting explained to me. "A near-infrared telescope will probably go up first—something that can see water vapor, oxygen, and not much else. Maybe it will find that stuff in the atmosphere of a nearby planet in the habitable zone. From there you can go to thermal-infrared and look for methane, which you might not find unless it's present in relatively high concentrations with the oxygen, as it might have been on Earth during much of the Proterozoic. Even if you do find it, there will still be ambiguities, and you probably won't convince everyone at first. Another possibility would be finding planets like the Archean Earth: you wouldn't see oxygen, but if you looked in the thermal-infrared, you'd probably see a lot of methane and maybe some organic hazes. The naysayers would kick and scream about that, because it seems to be much easier to abiotically build up significant concentrations of methane than oxygen—a dead planet with lots of volcanic activity and much more igneous, reduced ultramafic rock at its surface could do it. You'd have to start looking for other potential biosignatures that are much more difficult to detect, gases like nitrous oxide or dimethyl sulfide. For any interesting planets we'd find at first, there could be a whole series of follow-up missions done at greater and greater expense of time and money to nail down what exactly is being seen. It could go on for fifty years, a century, who knows.

"So the real question is," Kasting continued, "if the first mission does find oxygen in some planet's atmosphere, can that alone be persuasive enough to drive investment in the rest of this process? It's unquestionable that, on Earth, the rise of oxygen was the most fundamental change in our planet's history, because it paved the way for the evolution of complex life, of us. But on other planets, we could be fooled if we aren't careful."

Kasting had come up with two plausible ways a lifeless planet might masquerade as a living, oxygenated world for future space telescopes. The first probably played out in the early history of our own solar system, when Venus lost its water to a runaway greenhouse: as the hydrogen from water escaped into space, it should have left behind an ocean's worth of free oxygen that would have gradually reacted with carbon to form CO_2, leaving scorched Venus with an oxygen-rich atmosphere for perhaps hundreds of millions of years. This "false positive" didn't worry Kasting very much—the truth could be revealed by such a planet's location near the habitable zone's inner edge, and by the lack of any water vapor accompanying the oxygen in its near-infrared atmospheric spectrum. Kasting's second scenario was more troublesome, and involved a small, frozen planet at the outer edge of the habitable zone: if the planet was between perhaps two and three times the mass of Mars, it would likely be too small to long retain the internal heat that drives volcanoes and anything like plate tectonics, but large enough to prevent a thick atmosphere from being stripped away by stellar winds. Ultraviolet photolysis of even small amounts of water vapor would produce infinitesimal quantities of free oxygen in the upper atmosphere of such a "super-Mars," Kasting said; but without any volcanic gases to react with, and with surface water locked in ice and unable to expose oxygen-absorbing minerals in rocks, that slow trickle of oxygen could build up over billions of years to fill the atmosphere and give the illusion of life. Interpreting possible biosignatures grew even more fraught and uncertain when considering planets around stars significantly different from our own Sun—some red dwarfs, for instance, though smaller and cooler than our star, emit significantly more ultraviolet radiation, enough to radically alter the atmospheric photochemistry of the planets in their habitable zones.

As we drove out of State College along U.S. Route 322, bound for Black Moshannon, we passed across Bald Eagle Ridge, an Appalachian spur of quartzite, sandstone, and shale. The road forked, cut through the ridge, and split off into an extension of Interstate 99, a new stretch of highway that had been built in the first decade of the 2000s. As we continued on Route 322, running alongside I-99, I noticed that the long, sloping hillsides surrounding the roadcut looked strangely slick and denuded. I realized that I wasn't looking at rock and soil at all, but at sheets of thick gray and black plastic held in place by wire mesh. I pointed and asked Kasting if he could tell me what I was witnessing.

"There was bad acid runoff here a while back. When they made the roadcut, the Pennsylvania Department of Transportation dug right through the sandstone and used all the leftover pulverized rock as base and fill. The sandstone was laced with veins of pyrite, fool's gold, that the surveyors somehow missed in their rush to get this thing built." Kasting shook his head. "You could call pyrite a 'reduced' rock. It's made of iron and sulfide, so when you expose it to oxygen it breaks down into iron oxides and sulfates. Mix that with rainwater, which naturally contains carbonic acid from dissolved atmospheric CO_2, and the carbonic acid reacts with the sulfates to make concentrated sulfuric acid, which corrodes rock and leaches heavy metals from the ground. They started getting nasty runoff almost immediately from this whole stretch, and it was flowing into the groundwater and some good trout streams, so the highway got delayed by four years and many tens of millions of dollars. They had to go back and dig out a million cubic meters of this stuff and plant it in a landfill; then they covered up the rest. They could've saved themselves a lot of trouble by paying just a bit more attention, don't you think?"

If there is a theme underlying Kasting's work, it must be that having the patience to deeply ponder seemingly simple interactions of air, rock, water, and sunlight can yield surprising and sometimes profound insights. In fact, Kasting's deep patience was exactly what led to the greatest breakthrough of his career—an insight that revolutionized all subsequent studies of planetary habitability. It occurred to him one day in late 1979, while defending his PhD thesis at the University of Michigan, a few years before his father would urge him to "get a real job."

Kasting's breakthrough idea concerned how exactly the Earth had maintained a relatively narrow range of clement surface temperatures for billions of years despite the Sun's slow, steady brightening over that time. When our planet was freshly formed, the Sun should have been some 30 percent less luminous than it is today—a diminution more than sufficient to completely freeze the Earth's surface for the entire first half of its history. And yet scientists have found abundant evidence for liquid water on Earth throughout that time. Though researchers understood the stellar astrophysics underlying this "faint young Sun problem" in the 1950s, it was not until 1972 that planetary scientists became widely aware of it, via a paper by Carl Sagan and his colleague George Mullen. After that paper, prior estimates of habitable zones were left in disarray.

Reconstructive efforts began in the late 1970s, when an astrophysicist named Michael Hart at NASA's Goddard Space Flight Center simulated the faint young Sun's effects on the evolution of Earth's atmosphere and climate. Hart found his virtual Earth would only survive and evolve to resemble our current world if the inventory of greenhouse gases in its early atmosphere was significantly boosted. This came as no surprise—most researchers believed (and still do) that this was essentially how the early Earth avoided freezing. But Hart's other findings were more disturbing: If he moved the Earth 5 percent closer to the dimmer Sun, the enhanced greenhouse effect rapidly boiled the planet's oceans. Worse, if he moved the Earth only 1 percent farther away from the faint young Sun, once the planet acquired oxygen two billion years into its life, the subsequent decrease in greenhouse gases such as methane created bright glaciers that spread all the way down to the equator, reflecting ever more sunlight in an "ice albedo" feedback loop that ended with the entire ocean frozen solid. No matter how long Hart's model ran, the frozen planet remained forever locked in ice. No evidence had yet been found for our planet's own Snowball Earth episodes and its eventual recovery, so Hart believed the runaway glaciation to be an inescapably fatal problem. Hart's solar habitable zone was minuscule, and only by the rarest happy chance could Earth have formed in its midst. He chillingly concluded that there were far fewer habitable planets in our galaxy than had previously been thought. By Hart's estimation, Earth could well have been the only one.

James Lovelock had a wildly different idea. He believed that Earth had indeed endured the faint young Sun through some potent atmospheric mixture of greenhouse gases, probably mostly CO_2. But he posited that the reason our planet had avoided a runaway greenhouse early in its life was that photosynthetic organisms had pulled the excess CO_2 out of the air and locked it away in buried organic carbon at precisely the right rate to stabilize Earth's temperature. In his view, it was life itself that actively, unconsciously maintained the Earth's habitability by closely coupling and coevolving with the world's geophysical systems. The coupling was so close, he argued, that at the largest scales differences between living things and their inanimate environs became indistinct, and the world could rightly be viewed as a complex system analogous to a planetary-scale organism. He called this union of the biosphere and the rest of the Earth "Gaia" after the goddess of Mother Earth in Greek mythology. With a collaborator, the American biologist Lynn Margulis, Lovelock went on to author a large body of literature further developing the theory.

Kasting's contribution to this debate came from his study of carbon cycles for his PhD thesis, which was on the rise of oxygen on the prebiotic Earth. Specifically, Kasting was examining whether photolysis of CO_2 could have pumped significant amounts of oxygen into the atmosphere long before the advent of cyanobacteria and oxygenic photosynthesis. To tackle the problem, he first needed to estimate how much CO_2 had existed upon the primordial Earth, then feed that information into one of his custom-made numerical models. Much of the CO_2 present in Earth's atmosphere today is regulated by the biosphere, in an "organic" carbon cycle in which living things sequester carbon as they grow, only to release it back into the environment when they die and decay. But there is an older, inorganic carbon cycle as well, the carbonate-silicate cycle, one that operated before life had taken hold on Earth and still operates to this day on roughly million-year timescales. Kasting and I had glimpsed a small, isolated component of this inorganic cycle earlier, as we passed the roadcut with its acid runoff.

The inorganic carbon cycle begins when volcanoes belch CO_2

into the air, some of which then mixes with rainwater to fall as carbonic acid. On land, the carbonic acid weathers and erodes silicate
rocks, releasing carbon-rich minerals that accumulate in groundwater,
streams, and rivers. We had witnessed this first step as we crossed Bald
Eagle Ridge; most of the cycle's subsequent steps took place beyond
the spatial and temporal boundaries of human life. The carbon washes
into the ocean and eventually falls to the seafloor, forming layers of
carbonate rock such as limestone. When the actions of plate tectonics
push the carbonate-loaded seafloor down into the Earth's mantle, the
carbon cooks out of the rocks, generating CO_2 that streams back into
the atmosphere through erupting volcanoes, closing the cyclic loop.
While working on his PhD thesis, Kasting gathered the best estimates
he could find to constrain the early Earth's abiotic carbonate-silicate
cycle, and ran all the data through his model. At the end of the exer-
cise, he concluded that photolyzed CO_2 could have perhaps formed a
tenuous layer of stratospheric ozone, but little else, and certainly not
enough oxygen to enrich the atmosphere.

Kasting's thesis leaned heavily on the ideas of James Walker, an em-
inent atmospheric scientist at the University of Michigan who had taken
Kasting under his wing. Consequently, Walker sat on the panel that reviewed Kasting's thesis. Kasting successfully defended his thesis before
the review panel, and afterward the newly minted Dr. Kasting sat down
for lunch with his former interrogators. Kasting, Walker, and another
panelist, the atmospheric scientist Paul Hays, began discussing Hart's
troubling results and potential solutions to runaway glaciation on the
early Earth. Lovelock's theory seemed plausible but remained frustrat-
ingly vague and tautological: the Gaia hypothesis suggested that for a
planet to be habitable, it must first be inhabited. Perhaps, Walker offered,
something independent of life had acted to circumvent or counteract the
runaway glaciation—a dearth of clouds on the early Earth, for instance,
could have allowed more ice-melting sunlight to reach the planet's surface, or volcanic eruptions could have over time covered a glaciated
planet in a dark blanket of ash that absorbed more sunlight and melted

the ice. But these explanations felt unsatisfactory—whether or not they would occur seemed more a matter of chance than necessity.

With all the details of the inorganic carbonate-silicate cycle fresh in his mind from his thesis defense, Kasting paused for a moment to think, then said he had a simpler idea. "If the Earth were completely covered in ice, its interior would remain hot and volcanic activity would continue pumping CO_2 into the atmosphere," he haltingly began. "But there would be less exposed silicate rock, and the low temperatures would have frozen the water vapor out of the air. . . . So where could the CO_2 go? Why wouldn't it just keep building up in the atmosphere until the greenhouse effect melted the ice? Shouldn't weathering rates depend on temperature? Maybe that's the way out." Though they tried for the remainder of lunch, neither Walker nor Hays could find any objections to Kasting's observation. The next day Kasting left Michigan to begin a stint of postdoctoral research at the National Center for Atmospheric Research in Boulder, Colorado.

After taking us across Bald Eagle Ridge, Route 322 had dumped us in the adjacent valley, where we took another highway northeast toward the town of Philipsburg. Five miles of tree-covered ridgelines, grassy fields, and railway tracks passed by out the window, until Kasting motioned for me to turn left off the highway. As we began a looping ascent up a rough-paved road into rolling, oak-forested hills, he chuckled next to me in the passenger seat and with whimsy in his voice said, "That was probably the single best idea I've ever had, but I didn't even realize it at the time—I was still more interested in the rise of oxygen."

Ten months after his PhD defense, Kasting was working in Boulder when he received a large package in the mail. Inside was a thick manuscript titled "A Negative Feedback Mechanism for the Long-Term Stabilization of Earth's Temperature." Kasting was listed as the paper's third author, after Walker and Hays.

"Walker had gone off and worked out this whole thing," Kasting recalled. "Hays, I think, had helped him with some of the math. [Walker] had gathered all the information available about silicate weathering rates, looking mostly at lab data, and he pretty convincingly demonstrated that, yes, they do depend on temperature and rainfall. From all that data he derived an expression for the weathering rate as a function of CO_2 partial pressure and the temperature of the planet."

I asked Kasting to tell me again, in simple English, the crux of what the paper had said.

"It's pretty simple," he replied. "It said that when the temperature of Earth goes up, the rate of water evaporation increases, too. That puts more water vapor in the air, which 'takes up' more carbonic acid, which falls out in more frequent and intense rains. All that increases silicate weathering, which draws down CO_2 and cools the Earth. If the temperature goes down far enough to tip over into runaway glaciation, the buildup of CO_2 through decreased weathering provides a way to warm the planet again within tens of millions of years."

Kasting's voice had risen, and his hands had leapt from his lap to conduct the carbon-cycle symphony he held in his mind. "What we showed, what Walker showed, was that the carbonate-silicate cycle is just like a big thermostat, a stabilizing feedback that generally keeps an Earth-like planet's temperature away from dangerous tipping points. That's the whole key—the answer to Hart's problem, the abiotic alternative to Lovelock's Gaia hypothesis, the reason why the habitable zone is wide rather than narrow! Without this kind of stabilizing feedback, habitable planets would probably be as rare as Hart thought they were. With it, I can't help but think they must be very common."

After the paper's 1981 publication in the *Journal of Geophysical Research*, its core conclusions were quickly adopted by stunned planetary scientists around the globe. A separate trio of researchers—Robert Berner, Antonio Lasaga, and Robert Garrels—independently verified the paper's findings via a more complex study of the carbonate-silicate cycle, one partially based on measurements of dissolved minerals in rivers around the world. The new data revealed that rivers nearer the warm equator contained more of the carbon-rich minerals, while those in higher, colder latitudes contained less, in proportions consistent with Walker's derivation of temperature-dependent weathering rates. In the 1990s, geologists discovered the Snowball Earth episodes that occurred during the Proterozoic, further boosting acceptance of carbonate-silicate stabilization. In crust that had formed near the equator billions of years ago, they found layers of crushed rock that had been pulverized and deposited by glaciers. In equatorial beds of fine-grained deepwater marine sediment, they found dropstones—large, heavy rocks that had been plucked up and carried far from shore on the crumbling undersides of spreading glaciers. Glacial transport is the only plausible explanation for Proterozoic dropstones, for in that single-celled era of Earth's history no living creature existed that could hurl massive stones into the sea. Directly surmounting the ancient glacial deposits, geologists found the smoking-gun evidence for Kasting's proposed carbonate-silicate thermostat: hundred-meter-thick layers of warm-water carbonate rock, laid down in surges of photosynthetic productivity after an atmosphere saturated with volcanic CO_2 had rapidly melted away a shell of glacial ice.

In hindsight, the mechanism that Walker, Hays, and Kasting had discovered seemed as glaringly obvious as its applications. Suddenly, the differing fates of Venus, Earth, and Mars became much less mysterious. All seemed to have started with warm temperatures and liquid surface water, but only Earth had maintained those conditions, because only Earth had kept its carbonate-silicate thermostat. Venus lost its thermostat when it lost its water, since water is required to lubricate the motions of tectonic plates and to draw CO_2 from the atmosphere to form carbonate rock. Mars lost its thermostat not because it formed too far away from the Sun, but because it was too tiny. The planet ran out of geothermal heat to sustain the volcanism required to recycle carbonates, and its small size allowed most of the Martian atmosphere to slip away into space. Martian water that had once flowed in rivers and pooled in seas instead froze in the ground. If Mars had been a bit larger, it would have been able to recycle its carbon more easily, and it would probably still be habitable today.

The air had grown damp and cold in the shade of great oaks and black cherry trees, and only small patches of sunlight hit the road through occasional stands of tall, skinny pine. Ahead, Black Moshannon Lake stretched sinuously through a boggy clearing of sphagnum moss, evergreen sedges, rushes, grasses, and leatherleaf shrubs. Plant tannins tinted the lake's water the color of strong tea. Not another soul was anywhere in sight. We parked alongside a small man-made beach, and emerged beneath a cloudless blue sky. Kasting joked it was perfect weather for one of his 1-D models. "This place will be swarming with people next weekend," he said, glancing back to the forest's edge, ablaze with autumnal shades of crimson and gold. "We're only a few days ahead of peak color. Only a short while after that, maybe a week or so, the leaves will start to fall."

Prior to the revelation of the carbonate-silicate thermostat, astronomers had generally pegged the end of the world occurring some five billion years in the future, when the Sun will balloon to become a red giant that reduces Earth to a cinder. Planetary scientists had reasoned that, while the planet would still indeed exist by that far-future time, without oceans it would already be long dead—and so the end of the world could be set somewhere between one and two billion years hence, when the oceans boiled off into space beneath the light of a brighter Sun. The carbonate-silicate thermostat laid bare a new, more rapid route to the biosphere's demise: the gradual geologic drawdown of atmospheric CO_2. As the planet's interior slowly cools, volcanism will decrease, pumping less CO_2 into the atmosphere. Simultaneously, the steadily brightening Sun will be gradually raising temperatures, pumping more water vapor into the air to weather rock and draw down more CO_2. Eventually, atmospheric CO_2 levels will drop beyond the point where photosynthesis can occur, the base of the food chain will collapse, atmospheric oxygen levels will plummet, and the vast majority of life on Earth will die. Walker had realized this from the beginning, writing as the last sentence of the 1981 paper that "the terrestrial biota may, over the long term, have to adjust to the steady disappearance of carbon dioxide as well as the steady increase of average surface temperature."

In 1982, Lovelock and a colleague, Michael Whitfield, created an elaborate model of the carbonate-silicate thermostat to determine just how much time Earth's biosphere had left. Their results, published in *Nature*, estimated that doomsday would come in a mere hundred million years—a very short amount of time for a 4.5-billion-year-old planet. Translated into human terms, Lovelock's and Whitfield's prediction was equivalent to telling a forty-five-year-old woman that she has only one year left to live. Astronomers, planetary scientists, and geologists were shocked, but outside of these rarefied fields the news of the world's imminent demise largely fell on deaf ears—a hundred million years might as well be forever. Typical of dwellers in deep time, it would be another decade before scientists revisited the looming end of the world. In 1992, Kasting and his postdoctoral student Ken Caldeira performed a more nuanced calculation of Earth's photosynthetic decline, one that gave the biosphere a slight reprieve.

"Plants photosynthesize, but they also respire—they 'breathe' oxygen to help fix carbon in their bodies," Kasting explained as we approached the lake and began to walk along its shore. "Ninety-five percent of all the plant species on Earth, all the trees and most crops, almost everything, they rely on what's called 'C3' photosynthesis. The first step of their photosynthetic pathway makes a chain of organic carbon with three carbon molecules. If you get down below 150 parts per million, C3 plants have to respire faster than they can photosynthesize, so they die. In Lovelock's and Whitfield's model, atmospheric CO_2 hits 150 ppm in a hundred million years. Ken used my climate model, which is arguably better, and factored in the decay of organic matter and respiration of plant roots, which can pump up CO_2 levels twenty or thirty times greater in the soil than in the atmosphere. If you add that in, C3 plants can probably last five hundred million years."

Kasting bent down and ripped up a few green blades of grass from the soggy turf. "Lovelock and Whitfield also left out C4 photosynthesizers, which are more efficient with carbon. Grasses are C4. So are corn and sugarcane. They can subsist on only ten ppm CO_2. Our model showed CO_2 staying above ten ppm until about nine hundred million years from now, so maybe you'd lose trees and forests, but for another four hundred million years you'd have grasslands and cornfields. You'd have most of what's around this lake. C4 is a recent adaptation—maybe because of declining CO_2—so in all that time evolution might get even cleverer. But once you get below ten ppm, you're losing most of CO_2's greenhouse effect, so water vapor's positive feedback takes over, things heat up, the stratosphere becomes moist, and you lose all your water. All the CO_2 eventually comes back out of the rocks, but only after rising temperatures have cooked whatever's left of the biosphere. I wouldn't say our result is definitive, but it's using better assumptions, and it gives life on Earth another billion years instead of a hundred million."

"So Earth's biosphere is in its autumn years, in its decline," I prodded.

"I would say it's summertime for life on Earth, because lots of microbes can live at temperatures of 80, 100 degrees Celsius, which is how hot things will be when the planet starts losing water, and anaerobes and chemosynthesizers can persist even longer beneath the surface," Kasting matter-of-factly replied.

"Yeah, okay, the meek inherit the Earth. But what about for charismatic megafauna like us?"

"Maybe it's autumn for complex life. Let's just generously assume that humans or some form of intelligence can persist until C3 plants go extinct, and then we'll have real problems. That's five hundred million years out of what would be five billion years of detectable life on Earth. Intelligence could potentially exist here for one-tenth of the history of life on Earth, maybe one-fifth if you stretch things out with C4. The Cambrian Explosion was about five hundred million years ago. So maybe Earth has a billion, one and a half billion years of complex life in total."

Kasting stopped walking, and stood silently twisting the grass to shreds between his fingers. "I would say this bears on at least one of the terms in the Drake equation, the fraction of planets that develop intelligent life," he finally said, once again putting one foot in front of the other. "Whether through limitations of biology or in the geophysical evolution of the planetary environment, it took the first half of Earth's lifetime to develop complex life. Intelligence has only now appeared at the halfway point in the Sun's ten-billion-year lifetime, and it won't be easy to hang around longer than another half-billion years. That's a legitimate reason to think things like us are rare. People have called me an opponent of Lovelock's Gaia hypothesis because I helped find an abiotic way to stabilize climates, but I'm more of a critic. It's clear that life does alter its environment and can modulate climate to its benefit. It's also clear that life can throw the climate out of equilibrium. It's all a matter of perspective. Life caused the rise of oxygen, which probably caused runaway glaciation. That's not Gaian. But then again, the rise of oxygen led to us. It's probably a mistake to say there's any kind of purpose in this, but if Gaia could be said to have a purpose, the evolution of higher forms of life, of humans, could be it. And that's because humans in principle could postpone this planet's demise and extend Gaia's reach far beyond Earth. Intelligence and technology could prove to be more powerful than the cyanobacteria. You could call me a techno-Gaian. We probably can't prevent the Sun from getting brighter, but we could still protect the Earth. The Sun will be a problem in hundreds of millions of years, but if we maintain our progress, within only a century or two we should be in a position to counteract a brighter Sun by making some type of solar shield, maybe orbital clouds of little mirrors to block a fraction of the sunlight from hitting the planet. If we don't destroy ourselves or destroy the planet in other ways, we could protect the Earth for potentially billions of years. Why wouldn't we? We don't want to cook."

"You don't think we're destroying ourselves or the planet right now?" I asked. We had reached the lake's far shore without seeing another human being. With a sudden heavy roar, a white Ford F-150 pickup truck crested over an adjacent sloping gravel road that ran through the lakeside hummocks, its spinning tires pinging pebbles like buckshot into the trees and underbrush. Three startled cottontail rabbits broke from cover and bounded deeper into the woods.

Kasting frowned and tossed the torn grass on the ground. "I lose sleep over what we're doing as a species right now. It's not just the climate, either. We're squandering Earth's resources. We're doing terrible things to biodiversity. I have no doubt we're living in the midst of another major mass extinction of our own making. I take what little comfort I can from knowing we probably can't drive life itself to extinction or push the planet into a runaway greenhouse. The carbonate-silicate cycle will erase the fossil-fuel pulse in a timescale of a million years, and then the long decline of atmospheric CO_2 will continue. If we knew better, we'd hoard all the oil and coal and gas for when the planet really needed it. There's easily enough fossil fuel for us to raise the planet's temperature by ten degrees Celsius and make the Earth as hot as it's ever been in the past hundred million years, maybe longer. We could probably make the Earth the warmest it's been since the Archean. That would melt the ice caps, and we might lose twenty percent of continental land area to rising seas. Equatorial regions could become essentially uninhabitable, because many agricultural crops there are already quite near their heat-tolerance limits. Half the world's population could be displaced. Populations would shrink, and move poleward. Billions of human lives would be lost. . . . But technology keeps progressing. Maybe the global economy will recover in twenty, thirty years. Maybe we'll figure out reasonable ways to reverse or counteract some of the worst effects of climate change. Maybe we will end up building and launching a TPF, and whatever it finds will make us better appreciate our own planet. I think there's still time."

CHAPTER 8

Aberrations of the Light

The sky was hazy and overcast above Cape Canaveral, Florida, on the morning of July 8, 2011. A light breeze from offshore was the only respite against the sticky summer heat for the estimated 750,000 people who lined the beaches and coastal causeways surrounding Kennedy Space Center. They were there to say goodbye, awaiting the launch of NASA's space shuttle *Atlantis* into low Earth orbit and into history as it embarked on the final flight of the thirty-year-old space shuttle program.

As the final countdown commenced, the shuttle's last commander, Navy captain Chris Ferguson, contemplated the program's end with the mission's launch director, Mike Leinbach. "The shuttle is always going to be a reflection of what a great nation can do when it dares to be bold and commits to follow-through," Ferguson radioed from his seat astride the eighteen-story-tall, 4.5-million-pound shuttle stack—the orbiter, side-mounted to a hulking rust-colored external fuel tank and flanked by twin white rocket boosters. "We're not ending the journey today, Mike, we're completing a chapter in a journey that will never end."

Ferguson, like many before him, was echoing the core sentiment of the reclusive philosopher Konstantin Tsiolkovsky, the father of modern rocket science, who, from a remote log cabin in fin de siècle Russia, wrote passionately about space exploration and a human destiny amid the stars. Around the time that Orville and Wilbur Wright were pioneering powered flight at Kitty Hawk, on the other side of the planet Tsiolkovsky was theorizing about launching multistage rockets into orbit, living and working in outer space, and someday escaping the solar system. He famously devised what is now known as the "rocket equation," a single mathematical formula that encapsulates all the key variables affecting a rocket's maneuvers. Later luminaries like Germany's Wernher von Braun and Russia's Sergei Korolev cited Tsiolkovsky as influential in their pursuit of rocketry to explore and expand into space. In one of his early papers, Tsiolkovsky laid out the visionary impetus behind his work: "The bulk of mankind will probably never perish, but will just keep moving from sun to sun as each becomes extinguished. . . . There is thus no end to the life, evolution, and improvement of mankind. Man will progress forever. And if this be so, he must surely achieve immortality." Often implied but rarely explicitly acknowledged today for fear of cynical ridicule, this vision of an unbounded future beyond Earth remains the purest and most noble purpose behind any human space program.

"I still have dreams in which I fly up to the stars in my machine . . . ," Tsiolkovsky reminisced a decade before his death. "It should be possible to go into space with such devices, and perhaps to set up living facilities beyond the atmosphere of the Earth. It is likely to be hundreds of years before this is achieved and man spreads out not only over the face of the

Earth, but over the whole universe." Because of the seeming remoteness of his visions from the world in which he lived, Tsiolkovsky considered himself almost a failure, writing when he was sixty-eight that "I have not accomplished much and have not had any notable success." Tsiolkovsky died in 1935, still believing the conquest of space to be centuries in the future. Had it not been for World War II, he might have been right. But just over two decades after his death, Russian and American satellites were orbiting the Earth, the progeny of military spending on nuclear warheads and ballistic missiles.

Moments after Ferguson channeled Tsiolkovsky at Cape Canaveral, the shuttle's engines and rocket boosters surged and crackled to life, pushing *Atlantis* and her crew skyward on a quivering cataract of golden flame and electric-blue shock diamonds. The thunderous roar of its ascent swept over the surrounding scrubby marsh flats for one last time, attenuating over distance so that the farthest onlookers saw the shuttle launch in ghostly silence. *Atlantis* rose above the launchpad, rolled into an arcing trajectory bound for the International Space Station (ISS), and slipped beyond sight through the low veil of clouds. As the shuttle soared on a final flight into the heavens, it left behind a space program fallen into quiet decay: NASA lacked a ready replacement for the shuttle fleet and was scaling back programs across the entire agency. For years to come it would possess no direct capability to launch humans into space, and its science missions would be in retrenchment.

The shuttles had been conceived in the technological ferment of the successful Apollo missions to the Moon. As part of a larger twenty-year plan to build lunar outposts and send humans to explore Mars, NASA's top brass had pushed for funding to develop spacecraft that could launch like a rocket, rendezvous in orbit with robotic satellites and crewed space stations, then glide back to Earth through the fires of reentry to land like a plane at spaceports around the world. Unlike the gargantuan *Saturn V* rockets that were used once then discarded on the way to the Moon, such a system would in theory be fully reusable, creating an economy of scale to reduce launch costs, which at the time were higher than $10,000 per kilogram. The space shuttle was sold as a revolutionary spacecraft that would fly again and again, perhaps once a week, making space travel cheap, frequent, and routine. The wide, wondrous expanse of the solar system would be thrown open to human curiosity and ingenuity. Lunar bases and human explorations of Mars would be only the beginning of an incredible journey to the stars.

Instead, President Richard Nixon balked at the projected expense of NASA's grand plans. To ensure the agency would not pursue bases on the Moon and bootprints on Mars, he dramatically cut its funding and scrapped the Saturn rocket program. Perhaps because its focus was on making space travel more economical, the shuttle was the only piece of NASA's plan to survive, albeit in diminished form. Its goal of full reusability was scaled down to a "semi-reusable" design that was cheaper to develop but more expensive to operate. Even so, the shuttle's development required more money than a reduced NASA could provide, money that NASA found by appealing to the military's need for launching and intercepting spy satellites. In exchange for its support, the Pentagon insisted on changes—a more spacious cargo bay, a heavier thermal protection system, and larger delta-shaped wings—all of which ramped up the shuttle's complexity, cost, and risk.

The chimeric vehicles that finally emerged were elegant, versatile, and irreparably flawed. Instead of achieving 50 flights per year as originally projected, the entire shuttle fleet collectively flew 135 times during the program's thirty-year lifetime. The shuttles lofted payloads to orbit at a cost estimated anywhere between $18,000 and $60,000 per kilogram—more expensive than the expendable launchers they were built to replace. The shuttle program's failures came in part from the fact that many of its "reusable" components required extensive refurbishment by a small standing army of technicians after each flight. They also came from the shuttle's inescapable operational risks, which led to the tragic losses of two orbiters and crews. Space shuttle *Challenger* exploded shortly after launch in 1986 due to a sealant failure in one of its boosters, and the *Columbia* disintegrated during reentry in 2003 after a piece of foam insulation punctured a wing. Politically driven compromises made early in the shuttle design process proved to be major factors in both disasters.

The total cost of the program has been estimated at $150 billion, with a similar amount spent on its signature achievement, the ISS, a massive orbital laboratory that the vast majority of Earth's scientists did not want and could not use. For a time, the shuttles and the ISS collectively consumed nearly half of NASA's total budget, all while offering only the slimmest fraction of scientific returns in comparison to drastically less expensive robotic exploration. The most useful research to emerge directly from NASA's shuttle-era human spaceflight program only involved astronauts as test subjects, in experiments measuring the effects of prolonged space flight on human beings. Of course, the value of such research vastly diminishes without a capability to visit new destinations and perform meaningful work once off-world. Saddled with the financial strain of operating the shuttles and building the ISS, NASA saw its once-bold vision for a human future in space become a literal path to nowhere, in which astronauts looped endlessly around low Earth orbit waiting for their bones and muscles to waste away in microgravity. In nearly every aspect, the shuttle program was a ruinous white elephant that failed to deliver on its most crucial promises.

The lone exception was arguably the shuttle program's role in the Hubble Space Telescope, a school bus–size robotic observatory that was delivered into low Earth orbit by the space shuttle *Discovery* in 1990. After initially being proposed in the 1940s by the American astronomer Lyman Spitzer, the Hubble had been conceived and funded around the same time as the shuttles, and had taken decades and upwards of $2 billion to construct and launch. Its snug fit in a shuttle's cargo bay was not coincidental, as the telescope's design was derived from some of the same spy satellites the shuttle had been built to carry. It was not the first space telescope, but it was for its time by far the largest, with a 2.4-meter primary mirror of aluminized, precision-polished ultra-low-expansion glass—an almost eight-foot-wide eye above the sky. Above the Earth's atmosphere, Hubble's big eye would not have to contend with shimmering layers of turbulent air that muddied and distorted celestial light. It promised to revolutionize all of astronomy with the unheralded clarity and sensitivity of its observations.

Except, once in orbit, Hubble began beaming back blurry images. Its mirror had been polished into an extremely precise but ever-so-slightly incorrect figure, so that it deviated from ideal curvature by some 2 microns, less than one-third the width of a human red blood cell. The polishing had been a laborious two-year process that could not be readily replicated in orbit, and there was no way to swap out Hubble's mirror for a new one. Orbiting more than 550 kilometers (342 miles) overhead, the entire telescope was poised to be a multibillion-dollar boondoggle. That it instead became the most celebrated and productive observatory in history was due to the space shuttle's unique capabilities, and the good fortune that Hubble, uniquely among all space telescopes past and since, had been designed with astronautic upgrades and repairs in mind.

In December 1993, a solution devised by NASA's Hubble team was launched on the shuttle *Endeavour*. The space telescope would be given a set of smaller mirrors to refocus the starlight bouncing off the flawed primary mirror, rather like sliding a pair of glasses over myopic eyes. In the first-ever servicing mission of its kind, a crack team of seven astronauts clad in bulky spacesuits spent a total of thirty-five hours stretched across ten days installing the corrective mirrors and performing additional upgrades to Hubble. The marathon feat was rather like performing delicate eye surgery while wearing a welder's helmet and oven mitts—all in an unforgiving environment where even minor equipment failures could kill in an instant. Within weeks, the space telescope was delivering images unmatched by any ground-based observatory. Shuttles ferried new instruments and equipment to Hubble four more times during its life, with the telescope becoming even more powerful after each visit. Critics of NASA's human spaceflight program noted that for the estimated cost of each shuttle servicing mission, an entirely new Hubble could have been built and launched via expendable rockets, all without risking human lives, but even they couldn't complain about the new transformative vistas accessed by the shuttle-borne upgrades.

Within our solar system, Hubble's view was sharp enough to discern changing weather on Mars, explosive comet impacts on Jupiter, and Saturn's ghostly polar aurorae. Targeting Pluto, its keen sight revealed new moons and crudely mapped the distant body's surface. Gazing out to nearby star-forming regions, the telescope spied young stars cocooned in swirling disks of gas and dust that were midway through the process of forming planets. Measuring the motion of our nearest spiral galaxy, Andromeda, Hubble conclusively showed that the galaxy would collide and merge with our Milky Way in about four billion years. Inspecting surrounding galaxies, it found that nearly all of them had supermassive black holes at their centers, each squeezing the equivalent of hundreds of millions of Suns into a space smaller than the breadth of the solar system. Practically anywhere it looked, Hubble took gorgeous pictures that enthralled the public and enabled profound discoveries. In September 2012, astronomers released a "deep field" image a decade in the making, generated from two million seconds of Hubble observations. From time to time, they had pointed Hubble at what looked to be a blank patch of sky empty of any celestial object. The patch was small, with a field of view less than what you'd get by gazing at the heavens through a soda straw. Hubble's scrutiny revealed thousands of far-distant galaxies strewn like gems within that seeming void: yellow ellipticals, blue spirals, and jumbled "irregulars" in tangled skeins of color. The oldest and most distant galaxies manifested as small ruby points that gave no hint of structure—they had sent their light streaming toward us only a half billion years after the Big Bang, from stars that burned out long before the birth of our solar system. When the ancient photons arrived at Hubble's mirror, each was some ten billion times fainter than a human eye could detect.

In the years that followed Hubble's launch and repair, NASA expanded on the space telescope's success by constructing three additional massive space-based "Great Observatories," each devoted to a different wavelength of light and built at an average cost of about a billion dollars each. There was Compton, gazing at gamma rays from explosions at the edge of the universe. There was Chandra, watching with X-ray eyes as massive stars detonated as supernovae and supermassive black holes fed on molecular gas clouds. There was Spitzer, capturing the birth of stars and measuring the atmospheres of large, hot, transiting exoplanets in infrared light. All but one were launched by shuttles, and each added its own subset of breakthrough discoveries to what scientists soon began calling a "golden age" of astronomy. In addition to NASA's four "flagship" space telescopes, the agency also built and launched an armada of smaller, more specialized observatories at costs of a few hundred million dollars apiece.

The high costs of the flagships and their accompanying fleet of smaller telescopes were still primarily caused by the staggering expense of reaching orbit—an expense that the shuttles had failed to reduce. A launch cost of tens of thousands of dollars per kilogram cascaded into greater expenditures in the design, fabrication, and testing of each telescope, which had to be made simultaneously lightweight and rugged, as reliably as technology could allow. At the time, the expenditures weren't so worrisome—this was the America of the mid-1990s, a robust post–Cold War hyperpower with low unemployment, high productivity, a nation headed toward a trillion-dollar federal surplus, with a GDP and stock market soaring into orbit. Projecting forward, NASA's leaders thought they saw a bright future in which the agency's budget would steadily rise year after year, allowing the construction of even more-ambitious space telescopes, as well as sample returns from Mars and, ultimately, the rekindling of human exploration beyond low Earth orbit. After Hubble reached the end of its useful life sometime in the first or second decade of the twenty-first century, it would be de-orbited into the Pacific Ocean, and a new, even more revolutionary observatory would take its place. Hubble's successor was announced in 1996 as the Next Generation Space Telescope before being renamed the James Webb Space Telescope (JWST) in 2002 in honor of the NASA administrator who had guided the agency in the glory days of Apollo. Its mission would be to fully unveil the universe's very first galaxies, the objects that manifested only as tiny red blobs in Hubble's deepest images. JWST would be only the beginning—the U.S. astronomy community rapidly made plans to pursue many additional big, ambitious space telescopes, like a hungry diner selecting not one but several gut-busting entrées from a menu.

Just as NASA threw its weight behind JWST, the field of exoplanetology began its meteoric rise. Astronomers could for the first time rationally discuss the possibility of finding other Earth-like planets, to considerable public interest and acclaim. Over interstellar distances, the planet hunters calculated, our world would appear slightly fainter than a typical galaxy in a Hubble deep field image. In theory, that would be something JWST could detect, and indeed the telescope would excel at imaging hot young gas-giant planets far from their stars. But in practice, habitable planets would lie far too near their much brighter host stars—the planned telescope would not possess the high dynamic range required to satiate planet hunters or their suddenly adoring public. Our own Earth, for instance, is some ten billion times fainter than our Sun in visible light—for every photon reflected off our planet out to space, our star blasts out ten billion more. In infrared, the contrast ratio becomes more favorable—at those wavelengths, the Sun is only some ten million times brighter than our world. Astronomers like to compare imaging another Earth around a Sun-like star to photographing a firefly hovering near a bright spotlight from a viewpoint thousands of miles away, but the simple reality is more powerful. To image a rocky planet around a star is to capture a dim fleck of dust practically hugging the cusp of a thermonuclear fireball, like photographing an unlit match adjacent to a detonating hydrogen bomb. To do that, you would have to somehow block out all those thermonuclear photons in their millions or billions so that even a single photon reflected from a planet could be seen. For nearly all the stars in the sky, the blurring interference of Earth's atmosphere ruled out such precise measurements from the ground—only a space-based observatory could deliver the light of any potentially habitable planets orbiting many other stars.

At an early 1996 meeting of the American Astronomical Society in San Antonio, Texas, shortly after Geoff Marcy unveiled his team's first discoveries of hot Jupiters, NASA's then-administrator, Dan Goldin, took the stage to present an alluring vision of what the agency would do immediately post-JWST to support the search for other living worlds. Goldin intended to reshape NASA's entire science program around astrobiology, with new life-finding space telescopes as the glittering centerpiece. "About ten years from now," he explained, the agency could be ready to launch a "Planet Finder," an observatory that would locate potentially habitable worlds and, via various starlight-suppression techniques, take their low-resolution pictures. It would look for atmospheric biosignatures in the spectrum of each small clump of planetary pixels. This was one of the earliest public mentions of what would become NASA's "Terrestrial Planet Finder" (TPF) mission concept. If a TPF found promising worlds around nearby stars, Goldin told his rapt audience, then "perhaps in twenty-five years" even more ambitious telescopes could be built that could image those planets "with a resolution to see ocean, clouds, continents and mountain ranges." Goldin laid out a not-too-distant future in which maps of alien Earths would grace the walls of school classrooms around the world, courtesy of America's wealth and ingenuity. Sometime in the twenty-first century, he later said, those revealed as living worlds could become prime targets for robotic interstellar probes. By Goldin's rosy estimates, a TPF might fly as early as 2006, serving as the predecessor to another observatory that would arrive in the early 2020s to begin practicing Rand McNally cartography on any nearby terrestrial exoplanets.

Unfortunately, JWST's development proved more difficult than planned. To image the earliest stars and galaxies, the telescope would need a much larger primary mirror than Hubble's, one optimized for the infrared wavelengths where molecular clouds, giant planets, and the earliest galaxies shine the brightest. It would also need to be cryogenically cooled so that its own internal heat would not wash out the fragile light of cosmic dawn. Finally, it could not operate in low Earth orbit, because the lightbulb-like glow of our planet at infrared wavelengths would contaminate the delicate observations. Over several years and iterations, a design was nailed down: JWST would possess a 6.5-meter mirror with nearly seven times Hubble's light-collecting area and would be located at a point of stability between our planet and the Sun, almost a million miles from Earth, some four times farther away than the Moon. Nearly every aspect of the telescope would require major new technologies. A multilayered "sunshield" as wide and long as a Boeing 737 jet would protect the telescope and its suite of custom-built, state-of-the-art instruments and detectors. The entire assembly would be far too large for any existing rocket, so for launch the observatory would be folded up like origami, like a butterfly in a chrysalis, before unfurling in the depths of space. To fold, JWST's mirror would be divided into eighteen adjustable gold-coated hexagons, each chiseled from featherweight, highly toxic beryllium metal.

Various international partners signed on to construct instruments or to provide a launch vehicle, but NASA would bear the brunt of the cost, which early estimates pegged at approximately $1.5 billion. Launch was tentatively scheduled for circa 2010. As the project's true complexity and scale became clear, cost estimates were revised ever upward, but little in the way of boosted funding appeared. Instead, money for JWST would have to come from NASA's other space-science programs. In the end, more than $2 billion would be required for technology development alone. JWST's schedule began to slip, ballooning the project's total cost and shifting more and more of the major expenditures into the future. By 2012, JWST's construction, testing, launch, and first five years of operation were estimated to cost nearly $9 billion, with a launch date no earlier than 2018.

JWST's birthing pains were exacerbated by repeated national and global economic calamities, culminating in the Great Recession that began in 2008, in which the U.S. government spent trillions of dollars to prevent the total collapse of its major banks and other financial institutions. NASA's budget, once projected to steadily grow, now was fortunate to simply stay flat, and even then did not keep up with monetary inflation. The trillion-dollar surplus built up in the 1990s under President Bill Clinton had become a multitrillion-dollar deficit in the 2000s under the tax cuts and runaway spending of his successor, President George W. Bush. After the *Columbia* shuttle disaster, Bush had delivered a bold new mandate to NASA that harked back to the agency's original post-Apollo plans: build new heavy-lift rockets, then use them to return to the Moon and to deliver humans to Mars. It would be called the Constellation program. Alas, Bush did not deliver sufficient funding or deep congressional support, and scarcely mentioned the program after his initial announcement. As was typical of so many government projects begun during Bush's administration, the only thing Constellation seemed to excel at was transferring billions of dollars of public, federal money into the coffers of well-connected private contractors who too often delivered precious little in return.

In 2006, NASA chose to siphon billions of dollars from its science budget to prop up Bush's failing plan, throwing JWST's development into disarray and dashing hopes for a prompt development and launch of a TPF, which was officially "deferred indefinitely." Not everyone mourned the loss—many astronomers not studying exoplanets had come to view the narrow focus and projected cost of a TPF as an almost existential threat to their less-glamorous subfields that also required space telescopes. Indeed, some had actively lobbied against it within influential study groups and planning committees.

After years of middling results and more than $10 billion in expenditures, Constellation was canceled in 2010 by President Barack Obama, but the damage to NASA's science programs had already been done. To adequately fund JWST, the agency was forced to scale back, delay, or cancel nearly all its other major next-generation astrophysics and planetary-science missions; if the observatory was to succeed, it could only do so at the great expense of effectively eliminating most of NASA's space-science portfolio. As the prior generation of aging space telescopes wore out and broke down one by one, it seemed likely that whenever JWST finally launched, it would find itself almost alone, gazing out to the edge of the universe, the beginning of time, in a realm suddenly emptied of other major U.S. observatories. For want of money and strong institutional support, a TPF seemed almost as far distant and out of reach as the stars themselves. Congress repeatedly threatened to defund JWST due to the program's incessant delays and overruns—there was a chance Hubble's replacement would not fly at all. Even if it did, the telescope's feasible lifetime was only ten years, after which its fuel reserves would be exhausted and its instruments degraded. Astronomers began murmuring that, with JWST, the golden era Hubble had initiated was perhaps drawing to a close.

The thought weighed heavy on John Grunsfeld, a jovial, mustachioed astrophysicist and NASA astronaut who had flown on five space shuttle missions, including three to visit the Hubble. The telescope's success had come in no small part from Grunsfeld's space-suited dexterity, which he exercised in a record-setting fifty-eight and a half hours of spacewalking across his three Hubble servicing missions. The press hailed Grunsfeld as a hero, and called him "Dr. Hubble." Riding the shuttle into orbit to repair the most productive space telescope in history, then using that same telescope to study binary pulsars and other exotic celestial phenomena, Grunsfeld directly experienced the powerful synergistic benefits that could exist between NASA's human and scientific space programs. He mulled over the hundreds of billions of dollars spent on the ISS and the shuttles, and the comparative sliver of funding required to sustain the golden age of space telescopes. He wondered how, as with the shuttle and the Great Observatories, NASA's brawny human exploration program might once again forge a powerful partnership with the agency's purely scientific side, to great mutual benefit. During 2003 and 2004 he had served as NASA's chief scientist and had helped develop science applications for Bush's Constellation program. Big rockets, it turns out, are just as useful for launching extremely large telescopes as they are for hurling astronauts to the Moon. Such a rocket could, for instance, launch JWST without the expensive hassle of segmenting and folding the mirror. It could conceivably make larger TPF-style observatories cheaper as well. But the planning had backfired when Constellation cast its hungry shadow over NASA's science budget.

After completing the final Hubble servicing mission, in early 2010 Grunsfeld left NASA to serve as deputy director of the Space Telescope Science Institute in Baltimore, Maryland, the pulsing nexus of operations for Hubble and, someday, for JWST. For nearly two years, he worked closely with the Institute's director, the astronomer Matt Mountain, laying the groundwork for a future TPF-style telescope that the Institute might someday manage, too. Their preferred in-house design was aptly named ATLAST, the Advanced Technology Large-Aperture Space Telescope, and was intended as an astronomical workhorse that would, among other things, deliver images of potentially habitable exoplanets. Dr. Hubble had become Dr. TPF, Dr. ATLAST.

In his new role, freed from the stagecraft of being a high-profile NASA public servant, Grunsfeld spoke ardently and at length, often unbidden, about the importance and value of building new observatories to find other worlds and seek out new life. Then, in late 2011, Grunsfeld's phone rang. It was a friend at NASA. The agency wanted him to come back, to serve as the associate administrator of its Science Mission Directorate—a role that would place Grunsfeld at the helm of the largest pure science budget in the world, albeit one struggling to meet all its myriad obligations. He accepted, and upon returning tamped down his past vocal advocacy for life-finding space telescopes in favor of a more carefully managed public persona that emphasized balance across all of NASA's science programs. There was no announcement of bold new funding to search for alien Earths, but Grunsfeld's closest friends and former confidants could not forget his former fervor. After nearly a year's worth of fruitless e-mail exchanges with NASA's press team attempting to secure an interview with Associate Administrator Grunsfeld for this book, I took solace in what Deputy Director Grunsfeld had freely told me once during an earlier interview.

"Hubble and Webb will probably leave us hanging on the question of whether there is life elsewhere in the universe," he said. "What we need in the next generation of great space telescopes—and what we can achieve—is the capability to observe the atmospheres and surface features of every single habitable planet around the nearest thousand stars. We could finally find out we're not alone. We could finally find other habitable worlds, each of which, in principle, could be visited by humans. That's the big picture, and I want to be able to convince the public and the Congress and the future administrations that it's worth investing in this next step." Clearly, Grunsfeld had read his Tsiolkovsky.

I visited the Space Telescope Science Institute on a cold, foggy morning in early 2012, less than a month after Grunsfeld had departed to take the reins of NASA's science programs. The Institute occupies a nondescript building of tinted glass and dun-colored brick on the Johns Hopkins University campus and employs some five hundred scientists, engineers, and support staff. In the director's office, amid glossy posters of stellar nurseries, scale models of space telescopes, and Hubble mementos flown back from orbit by the shuttle, Matt Mountain warmly shook my hand and, in keeping with his British upbringing, offered me tea. Mountain is middle-aged, a wry, owlish figure with searching eyes beneath a mop of sandy curls. His standard uniform—a dapper suit— looked baggy over a once-stout build made vestigial by "a ruthless doctor." He had become director of the Institute in 2005, after spending a few years as JWST's telescope scientist and a long, successful stint supervising the development, construction, and operation of the twin

8-meter Gemini infrared telescopes. When he spoke, it was with the brisk but calm cadence of someone well versed in elevator-pitching complex and expensive projects to impatient politicians and technocrats—powerful people whose busy schedules did not afford the luxury of long attention spans. Early on, he unpacked a well-polished quip.

"In the twenty-first century, the discovery of life around another star will probably be as important a step for mankind as Neil Armstrong's was for the twentieth," Mountain articulated in plummy tones. "Finding life that has independently formed in another place, regardless of its progression or not to intelligence, will be like putting Copernicus and Darwin into the same bottle and giving it a good shake. What happens then? You look in the bottle. Maybe you revolutionize the world. I think this is a legitimate direction for NASA."

He produced an iPad from his desk and summoned imagery to accompany his words. "By the time we get to 2020, Earth-mass planets in habitable zones will be boring, you know. Because we won't have a clue what we're really looking at unless we have a *spectrum*." He flashed an image of six boxes, each filled with a substantially different squiggly line. The lines were simulated atmospheric spectra of the Earth at six different points in its geological history, from a lifeless atmosphere of carbon dioxide and nitrogen, to an Archean atmosphere filled with biogenic methane, up through the planet's incremental oxygenation. "There's only been a very small period in Earth's history where the action looks like this," he said, tapping the box containing Earth's present-day spectral squiggle. "We know an awful lot about the nearby stars—where they are, how old they are. Most of them are actually younger than the Sun, so maybe an Earth there will look more like these earlier epochs. But to know, we will need spectra. And once you admit that, suddenly you're in trouble, because it means you need a significant aperture in space—a telescope with a big mirror.

"I'm going to really upset everyone and just say that suppressing starlight to one part in ten billion is just an engineering problem," he went on. "Let's just assume we are really smart and solve that. Well, we still need a high enough angular resolution to spatially separate a planet from its star. And we still have the problem that Earths are bloody faint, fainter than a galaxy in a Hubble deep field! We're talking about something so faint, you can almost count its individual photons off on your fingers as they arrive at the mirror. It takes time to build up a spectrum, but in any reasonable mission we can't spend more than a million seconds staring at one of these things, because the telescope will need to be used for other research, too, and we may need to search many stars to find what we're looking for."

"How many stars are we talking about here?" I asked.

"Ahh!" Mountain exclaimed, tapping at his iPad. The screen went black, then what looked like a slowly revolving cloud of sparkling ruby, topaz, and sapphire spun into view.

"These are all the stars within two hundred light-years of the Sun," he said, then tapped again. The rubies and sapphires disappeared, leaving behind only orange, yellow, and white orbs. "Here are all the stars like our Sun, the ones we think are most likely to be habitable. In a computer, we can put an Earth in the habitable zone around every single one of these and ask, 'How many can we see with a telescope of a given diameter?' A telescope's resolution scales as its diameter, and its light-gathering power, its collecting area, scales by the square of its diameter. You need both to find something small and faint, so you put those factors together, and the number of candidates you can observe scales as the cube of the telescope's diameter. With a 4-meter, you can get . . ." He tapped the screen, and nearly all the stars vanished, leaving behind about two dozen in a central core clustered about the Sun. "A few."

An 8-meter would bring hundreds within reach, Mountain said between taps. A diffuse shell of stars materialized around the small core. "With a 16-meter you get thousands." He tapped a final time. The spinning swarm was shining, most of its original stars restored.

"Keep in mind this assumes every one of the nearby stars has an Earth in the habitable zone. We now know from Kepler that's probably too optimistic, and that somewhere between one and three out of every ten stars probably has a potentially habitable planet. We of course have no idea how often life gets started out there. So the question is, do you feel lucky? If you're very lucky, you can build a 4-meter and get away with it, because one of the handful of stars you'll look at will give you what you want. But what if you're unlucky? If you don't find anything in the nearest ten, well, you're not sure what you've learned. You may have just been dealt a bum hand. If you look at the nearest thousand and come up empty, then for all practical purposes we're probably alone. To have any rational chance of an answer, you should really look at hundreds of stars, and the physics of doing that thrusts you into this realm of 8-meter, 16-meter telescopes."

I nodded, picturing a large, flat, silvery disk slewing and tipping between targets in deep space, slowly building spectra from accumulated trickles of photons. That seemed simple enough. Why not go big and make it 16 meters wide?

Later, I did the math and fully grasped the difficulty. A 16-meter mirror, more than 50 feet in diameter, would have slightly more surface area than a regulation singles tennis court—quite a lot to fold into a rocket. Even if it were fashioned from lightweight beryllium segments like JWST's, the mirror and its support structure alone, sans instrumentation, would weigh more than 45,000 kilograms—about 50 tons—somewhat heavier than the Apollo spacecraft that had required a governmental crash program and the world's biggest rocket to reach lunar orbit. Much of that weight would serve no other purpose than to ensure that the mirror's micron-scale figuring could endure the intense vibrations and g-forces of launch as well as the frigid vacuum and zero gravity of deep space. A mirror to find alien Earths would require at least one of three things, each quite expensive: a rocket even larger and more powerful than Apollo's *Saturn V*; piece-by-piece orbital assembly like that which built the ISS; or dramatic reductions in mirror weight, cost, and performance tolerances.

"If you asked me today how much this 8- to-16-meter telescope would cost, I'd tell you I haven't a clue," Mountain said after I did ask.

"I don't want to go there. But that's not the question you should be asking. The question is not how much does it cost, it's what technology do you need to make it affordable? We must avoid the past mistakes of some other communities, like the particle physicists. Their entire field slowed right down in the 1970s and '80s because it was asking for technologies that no one else wanted or needed. They had to build almost everything from scratch, component by component, and it got very expensive. To get this done, we need to look at our industrial base and stick with technologies that other people want."

During the Constellation program's heyday, NASA had wanted a hulking rocket, the *Ares V*, a rocket so large it would surpass even the *Saturn V*. Shortly after departing the Gemini Observatory to become director of the Institute, Mountain got a phone call from Phil Stahl, an optical physicist at NASA's Marshall Space Flight Center. Stahl was looking for science applications for the *Ares V*, and wanted to know how much the Gemini 8-meter mirrors weighed. Mountain recalled that he told Stahl the mirrors were roughly twenty tons apiece.

"Is that all?" Stahl asked.

"Is that all?!" Mountain was flabbergasted.

"Matt, the *Ares V* will be so damn big we could just put a Gemini in it," Stahl explained. Soon, the Institute was concocting plans for ATLAST.

"That was the 'holy shit' moment, when we understood major progress in space science could occur simply by NASA getting new rockets," Mountain recalled. "A Gemini mirror is big, stiff, and you can easily test it on the ground. It's a much simpler system than something lightweight and segmented like JWST. A Gemini mirror only costs $20 million! You can just pack the whole thing into the rocket, light up the blue touch paper, and suddenly you have an 8-meter in space!"

Mountain's optimism for the "big rocket" approach had dimmed, however, with Constellation's constant overruns and eventual cancellation. Out of the ashes, in 2011 Congress had assembled plans for a nearly identical rocket, the Space Launch System, but he was not convinced the replacement would ever fly. NASA optimistically estimated its development costs at nearly $20 billion. Operating the rocket over a program's lifetime would probably require tens upon tens of billions more, yet it would most likely launch only once per year. Critics called it the Senate Launch System. Like its predecessor, it appeared to be a pork-barrel project designed less for affordable orbital transit and more for delivering jobs to the districts of influential members of Congress. The nascent rocket's fate would be determined by politics, not science and engineering.

As Constellation fell into its death spiral, Mountain and others at the Institute had sought a cheaper, alternate path to big mirrors in space. Perhaps there were ready ways to make large telescopes lighter and less expensive while also preserving their ultraprecise figuring. On the ground, newer observatories had decades earlier abandoned thick, rigid, monolithic mirrors for ones that were thin, flexible, and segmented. The new mirrors were cheaper but also floppy and easily deformed by shifting winds, changing temperatures, and a telescope's pointing slews. The secret to their success was "active optics," arrays of computer-controlled actuators mounted on each mirror's back that performed what astronomers call "wavefront sensing and control." Monitoring waves of light as they propagated across the mirror, a computer could manipulate the actuators to change the mirror's shape and orientation, precisely counteracting any detected deformities. The JWST's segmented mirror of beryllium hexagons already incorporated active optics to a limited degree in its design, allowing for periodic on-orbit adjustments on a timescale of days or weeks. To make larger mirrors that would still fit within the size and weight limits of existing launch vehicles, mission planners would have to use flimsier, lighter materials that would develop deformities even in the vacuum of space, requiring constant correction from complex active optics systems.

Lightweight, flexible, large mirrors in space stabilized by active optics seemed to be a game-changing idea, except for one crucial detail:

they were unproven. To Mountain's knowledge, no one had ever flown such systems in space before. Developing and flight-testing the necessary technologies could prove very expensive and time-consuming, potentially eliminating the immediate benefit of using lighter mirrors. Mountain and his peers tempered their enthusiasm, until they noticed something curious: a shopping spree by major defense contractors such as Northrop Grumman and Lockheed Martin. In recent years the aerospace giants had bought up a string of smaller companies that specialized in manufacturing either lightweight mirrors or active optics systems.

"Astronomers aren't the only people interested in large space telescopes," Mountain told me later in his office. "We've been talking about NASA, but there is another, much more well-funded government agency which tends to look down rather than up." He was referring to the secretive U.S. National Reconnaissance Office, the NRO. He arched an eyebrow, and noted that the Hubble had been an offshoot of the NRO's once-classified "Keyhole" series of spy satellites. "I don't have the security clearances, and I'd like to keep it that way, but you don't need security clearances to calculate what aperture size is required if you want Hubble-style image quality and need to avoid someone hiding from you by taking out their watch to time the passage of a satellite overhead." In a geostationary orbit nearly 36,000 kilometers (22,000 miles) above the Earth's equator, a satellite would move at the same rate as the Earth turned, effectively hovering in the sky over a fixed point on the planet. To gain useful high-resolution images of the Earth from such altitudes, Mountain implied, would require a mirror on the order of 10 or 20 meters in size. Placed in geostationary orbit over three or four strategic geographic regions, such mirrors could serve as unblinking sentinels, constantly monitoring nearly the entirety of the Earth's surface.

From such a system, you could run, but you couldn't hide, Mountain intimated. The weight reductions possible through active optics would be "what lets you actually launch the bloody things" on existing rockets. In all likelihood, the technologies for active optics and lightweight mirrors in space were more mature than publicly known, and if eventually declassified could greatly benefit science and society. Mountain extolled the possible virtues: besides imaging alien Earths, the light-gathering power of an 8-meter or 16-meter mirror would revolutionize the rest of space-based astronomy, allowing astrophysicists to witness the formation of supermassive black holes and probe the cosmic distribution of dark matter. More generally, he said, large, cheap mirrors could also prove useful for beaming solar power to receiving stations on Earth, or for monitoring our own planet's changing atmosphere at the resolution of individual clouds to constrain weather forecasts and climate-change projections.

Some months after my discussion with Mountain, the NRO presented NASA with a smaller but still significant gift: two unused space telescopes and related hardware sitting in a restricted clean room in upstate New York. The NRO considered the telescopes obsolete, and rather than keep them indefinitely in storage, chose to offload them on the nation's struggling civil space agency. Each telescope was outfitted with a Hubble-size, Hubble-quality 2.4-meter primary mirror—suitable for a host of astronomical observations but too small to be of obvious use in characterizing potentially habitable exoplanets. NASA would need to spend money on launch vehicles and instrumentation to properly utilize the NRO observatories, but the gifts freed up at minimum hundreds of millions of dollars that the cash-strapped agency could, if it chose, devote to developing technologies for larger, life-finding telescopes. Whether NASA would actually make that choice, however, was far from guaranteed.

"One of the problems we have at the moment is that NASA has yet to make up its mind what it wants to be when it grows up," Mountain told me during our discussions. "It's still very much in this mode of boys and their toys, of big rockets as jobs programs. NASA needs a more enduring vision than that, but it can't change without consultation with Congress and the American people. Ultimately, going out and looking for life, whether around other stars or on other planets in our solar system, that's an infrastructure that can create powerful partnerships between the agency's human spaceflight and scientific sides! That same sort of partnership was why the Hubble mission was so successful. Hubble had no peer because we could visit and renew it."

As he spoke, Mountain gradually slipped into a more colloquial vernacular, as if leveling over a beer with a skeptical Texas congressman. "So, for example, NASA wants to go to Mars. Well, they ain't gonna get to Mars before 2030, right? So what else are the astronauts gonna do in the meantime? You're not gonna get people to Mars building things smaller, you gotta build 'em bigger. Big infrastructures in space for commercial, scientific, and defense applications—that's the future. Maybe the astronauts should get even better at assembling big systems in space. Maybe the agency should invest in robotically servicing these big structures. Maybe we should better leverage our investment in the International Space Station. Oh, and by the way, we've got a great idea you can do all of that with."

Conceived in collaboration with three NASA research centers, the Institute's idea was called OpTIIX, a convoluted acronym for the Optical Testbed and Integration on ISS eXperiment. Proposed for launch to the ISS as early as 2015, OpTIIX would be a low-cost, scalable platform for testing the assembly and active correction of a lightweight, flexible, segmented mirror in space. Its 1.5-meter primary mirror would be composed of six fully actuated 50-centimeter hexagonal segments, each manufactured from sheets of silicon carbide glazed with atom-thin layers of vaporized metal. Collected starlight would bounce from the primary mirror up to a smaller secondary mirror, then back down to a series of tertiary "fast steering" and "pickoff" mirrors that would compensate for jitter and channel the light to cameras for imaging and wavefront control. Star trackers and gyroscopes would work in tandem with a latticework of lasers beamed across the primary mirror to precisely point the telescope and maintain its optimum figure. Thanks to those technologies, OpTIIX would deliver clear images of stars and galaxies despite being bolted to the outside of the ISS, which at any given time would be jostling and bucking in sympathetic frequencies with its cargo of noisy, weighty, rambunctious astronauts. If necessary, the astronauts could perform spacewalks to repair or upgrade the system, but OpTIIX would be designed to allow fully robotic assembly and maintenance as its modular pieces were ferried up to orbit.

"We are at the limit of what the present paradigm of heavy launch vehicles and restricted folding geometries can do," Mountain said after a time, reverting to his professorial mode. He gazed out his office windows, past a windowsill lined with five framed pictures of his family. The morning's fog had boiled off beneath the weak winter sun, revealing a sere landscape of bare trees and dormant grass.

"Right now we build and test space telescopes on the ground, then fold them up to fit inside a rocket. Right now, without bigger rockets, you can't get much bigger telescopes. Things like OpTIIX could be the beginning of a scale-invariant process of building bigger and bigger space telescopes, because you take out all the extreme tolerances. If you think about it, for the ground we don't assemble and test a telescope in a big basement somewhere then cart it up to a mountaintop, do we? No, of course not. We assemble it piece by piece on the mountaintop on the assumption we can then bring its components into alignment. Technical concepts like active optics let you assemble, align, and upgrade your telescope right where it belongs—in space. You can imagine putting things together by robots or astronauts or some combination, and then you just keep on going. You keep on going. You could scale up your telescope almost to infinity."

I asked Mountain how likely he thought it was that this vision would come to pass. He furrowed his brow and dragged a hand over his jaw, producing a sound like dry, windblown leaves.

Americans could well choose to further divest from space science, he finally said, seeming to address his ghostly reflection in the glass windowpane. "The reality is, we're using federal money, and an awful lot of it. If that money leaves, it doesn't necessarily come back. It flows elsewhere, into other priorities, which I'm not necessarily objecting to.

But you can imagine the discontinuous change. A lot of the capability we've built up could go away quite quickly. On the other hand, an argument can be made that investing in science and technology, space science as one part of that, is precisely what has kept this country going and can keep it going into the future as other economies rise—China, India, Europe, and so on. The real issue for me is, is the position we're in part of a natural evolution, or is this just a lucky event?"

In Mountain's view, the golden age of Hubble and the other great observatories was a fortunate aberration, something just as much a product of geopolitics and economics as it was of pure technological development and scientific progress. Its genesis could be found in the formative events of the latter half of the twentieth century—the Baby Boom, the Cold War, the Space Race. Astronomers had harnessed that unlikely coalescence of opportunities to create for themselves an almost-mythical dreamtime, a radiant era in which the boundaries of technological capability slipped beyond the mundane realm of Earth, and the horizons of scientific discovery reached the edge of the known universe. And now, perhaps, it was all at an end.

"Lyman Spitzer came up with the idea of Hubble in 1947, and we finally got the Hubble launched in 1990," Mountain said. "But if we hadn't had the space shuttles, if we hadn't had the Department of Defense developing its spy satellites, making Hubble a reality would probably have taken several decades more. That's the sort of era we're returning to, in my opinion. We spent a fortune on Hubble, but it generated its own momentum. It gave us Compton and Chandra and Spitzer and some completely new technology. It gave us JWST, this amazing, huge cryogenic infrared telescope. That was the aberration, that was the Baby Boomers at work. And now they're going away, and we've spent almost every single penny we've got, and we've got a new generation facing this fundamental shift. It's hard. . . . What astronomers need to recognize is that once a project's budget reaches a billion dollars, it enters a whole new realm where other factors besides pure science come into play. The science becomes a necessary but not sufficient condition. That's really why you and I are talking like this right now." He turned away from the window to face me.

"Someone must explain that understanding how the Earth works in fine detail and mastering space technology is actually pretty good for everybody involved. Someone should say that finding life elsewhere could be a humbling experience that would be good for humanity as well. Maybe it could finally give us the kick in the pants we need to fully realize that we could screw up *everything* if we don't get our act together. When Galileo lifted that little telescope up to his eye, he didn't quite know what he was doing, but he unleashed a revolution. Maybe we're on the verge of another one. We are now beginning to appreciate the complexity of the Earth system, and we are faced with controlling that complexity. We are now realizing that biology and astrophysics are intimately linked. These are hard concepts, but we need to master them as a species to survive. Otherwise, you know, maybe we will find life out there that arose independently, but that would actually be really bad news. Think about it: if extraterrestrial life is everywhere, but sentience and technology are nowhere to be seen, that probably means societies like ours don't survive very long. Instead, they self-annihilate. If we master all this complexity, we don't have to be in that position. We should battle this push toward the small, this turning inward."

The Institute's OpTIIX initiative ran out of money in late 2012, just after successfully completing its preliminary design review. Without an estimated additional $125 million, it would never reach the ISS.

CHAPTER 9

The Order of the Null

In 1996, when NASA's administrator, Dan Goldin, unveiled the agency's plans for a future fleet of space telescopes to image Earth-like planets, the vision he laid out was largely based on a single study, the results of which were published under the title *A Road Map for the Exploration of Neighboring Planetary Systems*. Goldin had commissioned the study only months before the first discoveries of exoplanets around Sun-like stars, and in the aftermath of those announcements its findings took on new urgency. The study was multitiered, with three separate teams and more than a hundred outside experts offering consultation, but its overall lead was Charles Elachi, a planetary scientist and electrical engineer at the Caltech/NASA Jet Propulsion Laboratory in Pasadena, California.

Elachi was overseeing the Laboratory's space and Earth science programs at the time, and would later ascend to JPL's directorship. JPL is legendary in space-science circles as the NASA center most responsible for the agency's greatest robotic explorers—the Pioneer and Voyager probes, the Mars landers, rovers, and orbiters, the Galileo mission to Jupiter, the Cassini mission to Saturn, the Kepler mission, and many others had been designed, built, or managed by JPL. With the exoplanet boom looming, JPL and Elachi saw an opportunity for further prestige and growth: while the Space Telescope Science Institute operated NASA's space telescopes, JPL and its affiliates would develop and build them. If the new telescopes found any promising planets around nearby stars, JPL might even construct the first robotic probes sent voyaging to other worlds outside the solar system.

In many of their *Road Map* presentations, Elachi and his coauthors referenced images like the famous "Blue Marble" photograph of Earth, snapped from a distance of 45,000 kilometers by one of the astronauts of *Apollo 17* as they traveled to the Moon in 1972. The whole-hemisphere image reveals the entirety of Africa, covered with jungle, savannah, and desert, as well as the arid Arabian Peninsula and much of ice-covered Antarctica. Whorls and wisps of white cloud stand out against the deep blue seas, and a cyclone can be seen swirling in the Indian Ocean. By showing the Earth as a lonely and fragile oasis in space, the Blue Marble had helped galvanize the environmentalist movement of the 1970s, and it became one of the most widely distributed images in history. What sort of space telescope would it take, the *Road Map* teams wondered, to reveal such details about a world orbiting another star? Their calculations were sobering: obtaining a Blue-Marble-style optical-wavelength image of an Earth twin orbiting one of the Sun's nearest neighboring stars would require a single mirror—a "filled aperture"—at minimum some 5,000 kilometers, or 3,000 miles, in diameter. That is, a mirror roughly the same size as the continental United States. Barring humans suddenly developing the technological capability to somehow convert large asteroids into ultra-smooth polished mirrors, such gigantic filled apertures appeared forever out of reach. And even if such a large mirror could be made, the issue of suppressing the 10-billion-to-one glare of starlight loomed as another enormous technical challenge.

Fortunately, the laws of physics offered a single solution to both problems. When light is emitted from the surface of a star, reflected off the atmosphere of a planet, or absorbed by the material of a detector, it acts like a particle. But as it travels through interstellar space or across a telescope's mirrors, it behaves more like a wave. Instead of photons pinging against a mirror like drops of rain, imagine a continuous wavefront of light impacting and propagating across every square centimeter of a mirror's surface simultaneously. This wavelike nature of light allows a curious trick that astronomers call "interferometry": rather than building, say, a 10-meter mirror, a physics-savvy astronomer could simply place two 1-meter mirrors at a "baseline" of 10 meters apart, combining the light from each mirror to produce a single image with a 10-meter aperture's resolution. The wavefronts of light propagating from a far-distant source such as a star can equitably fall on any number of interlinked smaller mirrors as if they are a single larger aperture. Place a 1-meter mirror in Los Angeles and another in New York, then link and synchronize them via a computer-controlled beam combiner, and you've made an interferometric array with a baseline of 5,000 kilometers and the resolution of a continent-size mirror. Its light-gathering power, however, would still be equivalent to those two meter-size mirrors, and the array's synchronization would be stymied by the curvature and rotation of the Earth and the overlying atmosphere; gathering enough photons to construct a single high-resolution image of an exoplanet would be entirely infeasible. In deep space, however, an interferometer would be above the atmosphere and could stare uninterrupted by the passage of day or night. Freed from gravity and planetary curvature, in theory it could be made arbitrarily large, with any number of individual mirrors to boost its sensitivity and a baseline of any length to boost its resolution.

Furthermore, when astronomers recombined the disparate waves of light gathered by each mirror, they could align the light waves so that the wave troughs of one beam would precisely overlap with the crests of another beam, splashing against and annihilating each other like out-of-phase ripples on the surface of a pond. The destructive interference would form bands of dark shadow within a resulting image. The shadows would be dark enough, in fact, to null out the bright light of a star, allowing the dim twinkle of accompanying planets to be seen. Short of using the Sun itself as a gravitational lens, an interferometric array offered the greatest hope of obtaining a Blue Marble image of any exoplanet.

Elachi and his coauthors seized upon the interferometer concept for a TPF, and designed a mission optimized for observing in the infrared, where the star-planet contrast is only 10 million, compared with 10 billion in the optical. Four 1.5-meter cryogenically cooled mirrors on a linear boom forming a 75-meter baseline would operate beyond the orbit of Jupiter, where there is less leftover dust from our solar system's formation to scatter and corrupt the faint light from nearby stars. If the mission was to operate closer to Earth, each of its mirrors would need to be doubled in size to 3 meters to compensate for the greater density of primordial dust that exists closer to the Sun. TPF-I, as the general mission concept came to be called, would deliver no Blue Marble images of alien Earths, but it would be capable of taking "family portraits" of planetary systems around the nearest thousand stars, with each planet manifesting as a single pixel in the TPF-I's detectors. Measuring the color of the pixel would hint at whether a world was rocky, ocean-covered, or sheathed in a thick envelope of gas. Cracking its light into a spectrum would allow the detection of atmospheric carbon dioxide, water vapor, and the possible biosignatures of methane and oxygen. Tracking the pixel's fluctuating brightness over months and years could reveal the planet's bulk geography—the locations of its continents, oceans, and ice caps—as well as its seasons. The success of the *Road Map*'s interferometric mission would then pave the way for larger future interferometric arrays that would use formation flying and laser communication to achieve baselines of several thousands of kilometers, missions that could perhaps replicate the Apollo Blue Marble for habitable worlds orbiting other stars. To pave the way for TPF-I itself, a precursor mission called the Space Interferometry Mission (SIM) would be launched. As first conceived, SIM would string seven small mirrors across a large boom, providing as much as a 10-meter interferometric baseline, sufficient to survey more than a hundred nearby stars for the astrometric wobbles of any accompanying Earth-mass planets in their habitable zones.

Spurred by Goldin's enthusiasm and the tacit support of the Clinton administration, NASA quickly greenlighted SIM and convened working groups to solidify plans for TPF-I. Both projects eventually ran into major difficulties. Riding a strong initial pulse of funding, SIM met or surpassed all of its key developmental milestones, but by the mid-2000s the ballooning costs of JWST and of the Bush administration's new Constellation program had reduced the project's funding to a dribble. Most astronomers were unconcerned—SIM's hyperspecialization seemed to offer little to the broader community. Even many planet hunters thought it superfluous, and hoped to simply skip over it to build a much more capable TPF. The mission was repeatedly downgraded and its launch continually delayed, piling on empty expenses until, after consuming more than half a billion dollars, in 2010 SIM was quietly cancelled and its nearly complete flight hardware junked or repurposed.

TPF-I faced a different problem: As its working groups delved deeper into the related technological hurdles, they realized the initial estimates for the mission's cost and launch date were hopelessly optimistic. Cryogenically cooling all the separate mirrors would be costly and difficult. Reaction wheels required to rotate and point the mirrors on their long boom would cause the entire assembly to vibrate, potentially ruining observations. New designs emerged, including one from the European Space Agency's own TPF-I project, code-named "Darwin." Darwin and related concepts would eliminate vibrations by discarding the long boom in favor of a free-flying array of several mirrors that would gather light and direct it into a central beam-combining hub. Instead of one cryo-cooled spacecraft, the project would now require five or six, each needing to fly in formation with centimeter-scale precision in deep space, drastically increasing mission complexity and the amount of propellant required. The runaway cost growth of the complex, cryogenic JWST suggested that, if anything, TPF-I's early cost estimates of $1.5 billion would balloon to make it even more ruinously expensive than its predecessor. By 2001, JPL's notional launch date for a TPF-I had slipped to no earlier than 2014, and mission planners were looking for cheaper alternatives, ideally a single non-cryogenic telescope.

Conventional wisdom held that the very same slippery wavelike behavior of light that enabled interferometry would prevent any single filled-aperture telescope from ever imaging Earth-like exoplanets. To capture the 10-billion-to-one photons emanating from an alien Earth at optical wavelengths, the light must be strictly controlled, the star's overwhelming glare removed. Yet when starlight falls upon a single mirror it flows in liquescent wavelets, pooling and puddling in frozen ripples and coruscating speckles around the most minuscule surface imperfections. Even a mathematically perfect mirror, of the sort that only exists in computer simulations and the late-night dreams of theorists, would not be immune: light from a point-like distant star striking an ideal circular mirror would still diffract off the mirror edges, forming a central bright disk surrounded by a concentric series of rings. A good number of the disks, ripples, rings, and speckles tended to manifest precisely in the part of a star's image where one would expect to find any lurking habitable planets. Each aberration would typically only be about a hundredth as bright as a target star, but would still be some eight orders of magnitude brighter than the faint light of any accompanying small, rocky worlds, rendering planetary detections improbable, if not entirely impossible. This was the scientific consensus as presented by any up-to-date optics textbook at the turn of the twenty-first century. And it was totally wrong.

The key to a single-telescope TPF solution was a device called a coronagraph that could, in theory, blot out a star's diffraction disks and rings. Invented by the French astronomer Bernard Lyot in 1930 to observe the hot, nebulous corona that surrounds the Sun, a coronagraph is any occulting object placed in front of a telescope's mirror to block out the unwanted light of a target star. To see a coronagraph in action, make one of your own. Hold your right thumb over the Sun's disk in the sky to prevent most of its glare from reaching your eyes—the principle is the same. You may notice, however, that even if the Sun is entirely blocked, a small amount of sunlight still diffracts around your thumb's edge. You can dampen some of that extra glare by placing your left thumb a short distance directly behind your right thumb as an extra barrier to block the Sun from your line of sight. In his coronagraphs, Lyot did something similar, crafting a series of "pupil" lenses, partially transparent "masks," and disk-shaped opaque "stops" that progressively stripped out the residual light scattered off the edges of the initial occulter. Lyot's instruments were suitable for imaging the Sun's corona, which is a million times fainter than the Sun itself. But they leaked far too much stray light into a telescope's optics to allow for the crucial 10-billion-to-one starlight suppression required to image an exo-Earth in visible light.

In 2001, while pondering the mounting intricacies of TPF-I, the Harvard-Smithsonian astronomers Wesley Traub and Marc Kuchner struck upon a concept for a new class of coronagraphs, ones that more explicitly relied upon interferometric principles to suppress starlight. Traub and Kuchner found that by superimposing interferometric nulling patterns of spirals or bars on a coronagraphic mask and carefully tweaking the shape of a coronagraph's stop, they could simultaneously increase the total amount of starlight suppression to block 99.999999999 percent of a star's light while also channeling the residual starlight to a thin outer region, away from the coronagraph's central dark shadow. A star's light would be blocked, nulled out, and finally swept away to the edge of a detector, while the faint light of a nearby planet would pass through unimpeded to form an image within the shadow. The scheme worked almost flawlessly in tightly controlled laboratory tests. Traub and Kuchner's coronagraphs were straightforward to manufacture, but each mask typically worked best only for a few wavelengths of light rather than a star's full spectrum. Around the same time that Traub and Kuchner were working on their coronagraphs, another astronomer, David Spergel of Princeton University, independently devised a completely different arrangement of shaped coronagraphic masks and pupils that also achieved extreme starlight suppression.

JPL and NASA took note, and began funding research into a coronagraphic TPF, TPF-C, a planet-finding telescope meant to operate in optical rather than infrared light. Soon, a rough architecture had emerged: TPF-C would use a large 8-meter monolithic mirror mated to one or more specialized starlight-suppressing coronagraphs inside the telescope. The monolithic primary mirror would be oval-shaped rather than circular, so it could fit into a rocket fairing; segmented foldable mirrors like JWST's produced too many wavefront aberrations to be compatible with the ultrasensitive coronagraphs. In the aftermath of the breakthroughs by Traub, Kuchner, and Spergel, even more coronagraph designs emerged. Starlight could be suppressed in myriad ways, weakened in mazes of grooves, twisted in networks of spiraling vortices, or phased out and dispersed in labyrinths of masks and lenses. But all still leaked some fraction of unwanted light. The degree to which each design sprayed splashes of photons elsewhere in the telescope was called "the order of the null."

Strengthening the order of the null required piping perfect, flawlessly symmetrical wavefronts of light to and through the coronagraphs. The barest fraction of a guidance error in the telescope's pointing would cause the collimated beam of starlight to almost imperceptibly "walk" across new paths, wandering over different patterns of imperfections in the mirrors, weakening the order of the null. TPF-C would need to point with an accuracy more than five times that of the Hubble. A surface deviation of less than the diameter of a single silicon atom anywhere in the telescope's reflective surfaces would send imperfect wavefronts cascading down the optical train, weakening the order of the null. Compared with Hubble's, TPF-C's mirrors would need to be some hundred times smoother. Producing and maintaining such precise pointing and figuring would require significant and costly breakthroughs in vibrational control, active optics, and mirror manufacturing, but those tasks still seemed cheaper than any conceivable TPF-I mission. TPF-C would be less capable and sensitive than a TPF-I, and would survey fewer stars for potentially habitable worlds, but its probable lower costs made it a winner to budget planners at NASA and JPL. Year by year, TPF-C became ascendant, and TPF-I's fortunes declined.

Sensing the change in the winds, in 2005 JPL offered Traub a job, both as TPF-C's project scientist and as chief scientist overseeing NASA's exoplanet science programs. He would lead a team of approximately fifty people working to the tune of $50 million per year to do whatever it took to get TPF launched on time. Accepting the position would mean pulling up his stakes in Massachusetts and moving across the country. He was approaching seventy years of age, and knew if he left he would be uprooting old friendships as well. Still, Traub quickly made up his mind and took JPL's offer. He reasoned the new opportunity would be worth the sacrifice—within perhaps a decade, if all went according to plan, he would be leading TPF's team as observations streamed in, poring over the gathered light from other, distant habitable worlds. Of all the people to ever live—past, present, and future—perhaps Traub would be one of the lucky few to first find life beyond Earth, beyond the entire solar system. He could play a crucial role in the most profound development for humanity since the discovery of fire. Traub soon arrived in sunny Pasadena and settled into a small rented apartment.

As the fates of TPF-C and TPF-I diverged, a third method for suppressing starlight emerged, largely based on the work of Spergel, his Princeton colleague Jeremy Kasdin, and Webster Cash of the University of Colorado. All three researchers were concerned by the extreme tolerances required for TPF-C's mirrors. Rather than putting a corona-graph inside the telescope, inviting in all that corrupting, contaminating starlight, they proposed placing a coronagraph outside the telescope, as a separate free-flying spacecraft that would prevent any stray starlight from ever entering the optical train. They called the free-flying coronagraph a "starshade," and their simulations of its performance revealed that its ideal shape for diffracting and nullifying starlight would closely resemble a many-petaled sunflower. Unlike the starlight-suppression techniques of TPF-I or TPF-C, which required reams of custom-built kit and only oper-ated in a limited number of wavelengths, a starshade would simply cast a deep shadow onto a telescope, any telescope, allowing more broadband spectroscopy to widen the search for biosignatures. A starshade's telescope would not require cryogenic cooling like TPF-I, or ultra-smooth mono-lithic mirrors like TPF-C—any space observatory with a sufficiently large general-purpose mirror would do, including NASA's planned JWST.

But building and operating a starshade would be no easy task—many of the extreme tolerances associated with the TPF-C telescope would instead just be exported to a separate spacecraft. Most designs en-visioned a starshade between 50 and 100 meters in diameter, gossamer-thin and razor-sharp at its edges, sheathed in dark anti-reflective coatings, floating anywhere between 50,000 and 150,000 kilometers in front of a space telescope. For comparison, the distance between the Earth and the Moon averages some 380,000 kilometers; properly aligning a star-shade's shadow with a telescope would require exquisite orbital control. The starshade would have to autonomously unfurl in space and preserve its vast shape to submillimeter-scale precision, all while using small high-impulse thrusters to linger on or slew between targets. Where an agile TPF-C could switch targets in seconds or minutes, for a starshade the task would take days or weeks. A starshade would survey fewer stars than TPF-C, but at a potentially lower total cost. The starshade concept came to be called TPF-O ("O" stands for "occulter"), but its relatively late entry into serious consideration would relegate it for years to also-ran status among NASA's mission planners.

TPF-I, TPF-C, and TPF-O. After nearly a decade of heavy study, NASA, JPL, and other affiliated institutions had identified three broad technologies, each of which could produce images of habitable planets around other stars. All they lacked was a mandate and funding from NASA's political taskmasters to select an architecture and move forward. The mandate came, to much rejoicing, in a single statement buried in the supporting literature for President George W. Bush's 2004 vision for NASA, the vision that created the Constellation program to return astronauts to the Moon and onward to Mars.

Traub wistfully recalled the spirit of the moment when I spoke with him in the summer of 2012, long after the early promise of TPF had become a frozen dream. Traub is a tall, quiet man, with kind blue eyes contrasted by a blond coif and goatee progressively whitening with age. He was in the midst of moving to a new office at JPL, where he still served as head of NASA's Exoplanet Exploration Program. His desk was surrounded by blue filing boxes that, in aggregate, contained 285 linear feet of books, articles, and communications Traub had accumulated during his half-century scientific career. The bulk had accrued during the past seven years, during his JPL tenure, and much of it was related to the TPFs. He was cutting his stockpile down to 140 linear feet, filling trash bins with a good portion of the recent corpus on the telescopic search for extraterrestrial life. He pulled a folded piece of paper from one of the boxes and examined it through golden wire-rim glasses perched high on his nose. It was a memo from Charles Beichman, a Caltech astronomer who had been a key contributor to Elachi's *Road Map* and who was, in years past, the project scientist for TPF initiatives.

"This is from April of 2004, a year and two months before I came here," Traub said. "It's an unusually cheerful letter, considering Chas. He sent it to the members of the TPF science working group, of which I was a member." He cleared his throat and began to read: "'I want to inform you of exciting new developments for TPF. As part of the President's new vision for NASA, the agency has been directed by the President to,' quote, '"conduct advanced telescope searches for Earth-like

planets and habitable environments around other stars."' End quote."
Traub sighed softly and dropped the letter on his desk. "We've been living off that statement in the President's vision for NASA for more than
eight years now."

Emboldened by Bush's apparent support, NASA and JPL had
come to an audacious decision: Rather than choose between the infrared TPF-I and the optical TPF-C, the agency would fly them both, and
soon. NASA and JPL would build and launch TPF-C as soon as 2014,
then work with the European Space Agency to build and launch TPF-I
before 2020. Scientifically, the case for synergy was solid: spectroscopic
observations in both the optical and infrared would allow a far more
reliable determination of a planet's habitability and possible biosphere.
Beichman's 2004 memo served as the unofficial announcement, explaining that this was "the opportunity to move TPF forward as part of
the new NASA vision," and that "in the estimation of NASA HQ and
the project, the science, the technology, the political will, and the budgetary resources are in place to support this plan."

The planet hunters and the public were ecstatic, but many other
astronomers were resentful. NASA had chosen to build not one but two
expensive space telescopes devoted almost entirely to exoplanets, all
without ever officially consulting the various high-level committees
and study groups that tried to orchestrate national plans for space science. Building both TPFs, the critics argued, would leave no money
for other more worthy priorities, such as clarifying the nature of dark
energy, detecting gravitational waves, and observing active galactic nuclei in high-energy X-rays. In public, the pushback was muted, but privately it seethed. By the time Traub arrived at JPL in 2005, the howls of
acrimony had already begun to pull the Laboratory's lofty aspirations
back down to Earth.

"There is not a great deal of happiness among classical astronomers
with planets," Traub said with typical understatement. "Exoplanets are
even worse. Planets generally seem to be frowned upon by the astronomers who only look at stars and galaxies. There are a lot of people with

the attitude that it's fine to study astrophysics, the Big Bang, the evolution of galaxies, and the evolution of dust disks around stars. But don't ask whether the disks make planets. Don't dare wonder whether the planets make things that can hop and crawl around. Because that's somehow beneath our dignity, to think about things that might have anything to do with subjects as complicated as biology and life."

When Traub came to JPL, he encountered a stoic acceptance that it would not be easy to muster sufficient community support to fly one TPF mission in the very near future, let alone two. "There was a sense that, well, we certainly weren't going to be able to have two missions at the same time, and it wouldn't be happening as fast as this letter said," he explained. "But that was okay. We'd just work hard and get the science definition reports written, get the technology studies nailed down, and bring everyone onboard. It would take maybe a few extra years, but that was all. In retrospect I've learned, to my horror, that the science isn't everything. In fact, it's probably the last thing. People don't support what they think is best for all of science, they support what directly benefits them. These days, what the astronomy community pursues is full employment for astronomers."

As it became clear that building both TPFs would easily be multibillion-dollar endeavors, outspoken support for the project in the wider U.S. astronomical community became ever more muted. The choice for NASA became an easy one as it struggled to balance the budgetary burden of its human spaceflight efforts against the discordant demands of astronomers and astrophysicists. In February 2006, it cannibalized some $3 billion from its science budget to support a handful of space shuttle flights and Bush's Constellation program. The rapid push for life-finding space telescopes was effectively cancelled, officially demoted to become one of the agency's numerous and indefinite technology development projects, a trickle-funded limbo where grand visions went to die. Without a major shift in policy and funding for NASA, the agency seemed set on a course to delay any planet-imaging telescope worthy of being called TPF until, at the earliest, the mid-2030s.

In the meantime, the discoveries of more and more potentially Earth-like worlds would keep piling up. The transformative project Traub had uprooted his life for had, within less than a year of his arrival at JPL, crumbled to pieces. When I asked his opinion about TPF's prospects in his lifetime, Traub told me with thinly veiled gloom that remaining optimistic had become part of his job description.

"Professionally, I can't get discouraged, because otherwise this would be a very, very depressing thing to do," he said. "But when you start out to build a cathedral, it's not a requirement that you build it before you die. They usually took a couple of hundred years. We aren't building medieval cathedrals here, however. I think this is easier, and I do believe that if things had gone ahead as had been planned in that letter from 2004, if we had just been given the funds to go ahead, then we would be flying these telescopes more or less on the schedule that had been set. TPF-C would be nearing its launch, with TPF-I maybe at the end of the decade. There is nothing fundamentally new I have learned since 2004 to change that point of view. Now, technically speaking, none of this is easy. This is all hard compared to everything else I've ever done in my life, and everyone else working on this would tell you that, too. But right now, today, we do already have half a dozen ways to get the light of potentially habitable exoplanets. We've proved them in the lab. We need time to get the engineering correct, but not to invent new things. It's really just a matter of building a big mirror with a little piece of glass in the focal plane, putting it in a big rocket, and putting some deformable mirrors in the back!"

If it was so straightforward, I offered, maybe NASA wasn't the only answer. Maybe the solution could come from private funding rather than government sponsorship.

Traub shook his head. "There is nobody in the private sector with an incentive to spend money on something of this magnitude," he said. "It's almost impossible for people with money to spare to invest in a long-term project like this. That's why the government does it. NASA didn't decide to send men to the Moon; President Kennedy did. It's the legislators and the administration that tell NASA what to do, so what it's going to take is someone in Congress, someone in the presidency, feeling strongly about this and realizing that the first discovery of life beyond the solar system is an event that in all of history will only happen once. Do we want to be the ones who dropped the ball, who screwed up and didn't move this forward? All we need is for our political leadership to decide this is something important for NASA and for our nation to do. I can guarantee we know exactly how to proceed if we are given the go-ahead. That's my final thought on that."

I had first met Traub the previous year, in late May of 2011, at a small conference held on the campus of the Massachusetts Institute of Technology in Cambridge, on the glass-sheathed top floor of its famed Media Lab. Entitled "The Next 40 Years of Exoplanets," the conference had been conceived by the MIT astrophysicist and planetary scientist Sara Seager to mull over the field's troubled recent past and its possible future redemptions, redemptions that could come via TPF or in some other unimagined form. Seager had invited Traub to discuss the Kepler results and to defend JPL's role in TPF's rise and fall. She had invited many other luminaries as well. Matt Mountain had come to make the case for a poor man's TPF-O, explaining how a starshade could utilize less than a tenth of JWST's observing time to deliver spectra of any small, rocky worlds around a handful of neighboring stars. A JWST starshade, he estimated, would cost about $700 million, though NASA would be loathe to spend a penny more than what was already being budgeted to get its flagging space-science flagship into orbit. John Grunsfeld was there, too, seemingly already preparing for his return to NASA, hinting that America's astronauts were eager for challenging missions like assembling and servicing planet-finding telescopes in deep space far from Earth. His inner Tsiolkovsky emerged to declare that extinction awaited any single-planet species, and he optimistically predicted that solid proof of the first habitable exoplanet would come from a NASA space telescope on July 21, 2025—the fifty-sixth anniversary of humanity's first footsteps upon the Moon.

Seager was the conference's ideal catalyst. She was still relatively young, on the eve of her fortieth birthday, possessing sufficient passion and longevity to keep her at the forefront of exoplanet research for the next forty years. Though young, she was already one of the most respected and accomplished workers in the field. She had embarked on a career in astrophysics hoping to delve into cosmology, to reveal the formative early life of the universe. When the exoplanet boom began, she rapidly changed course. Beginning in the mid-1990s, when she was only a graduate student working under the Harvard astronomer Dimitar Sasselov, Seager had performed the first detailed theoretical modeling of the structure and evolution of hot Jupiters' atmospheres. At the time, many astronomers still thought hot Jupiters were illusory products of stellar variability and wishful thinking, and some viewed Seager and Sasselov's work as foolishly risky. Yet by 1999, she had obtained her PhD from Harvard, and the wider astronomy community had sheepishly caught up with her: most everyone finally agreed hot Jupiters were real, and Seager's models set gold standards for observational studies. In response, Seager surged ahead again, describing how a transiting hot Jupiter's atmosphere could be investigated without having to first build something akin to a TPF. In Seager's proposal, coauthored with Sasselov, she pointed out that starlight blasting through the planet's upper atmosphere would beam spectroscopic information toward Earth that astronomers could then discern using existing ground- and space-based telescopes; she recommended looking in particular for signs of sodium, which she calculated should project a clear spectroscopic signature in optical wavelengths. At the time, no transiting planets had yet been found. A couple of years later, a team tried out Seager's suggestion, using the Hubble Space Telescope to observe a newly discovered transiting hot Jupiter. As predicted, they found the spectral lines of sodium—the first detection of an exoplanet's atmosphere.

Through the years, Seager's focus had increasingly shifted to the search for exoplanetary life, in which she performed groundbreaking work on how to characterize the environments of potentially habitable worlds. She made it no secret that she hoped to lead any eventual TPF mission that flew in her lifetime.

Seager had organized the conference with an eye toward posterity, and had meticulously ensured that its proceedings were captured on video and archived online. She cut a slim, striking figure when she stood to deliver her opening remarks in front of the seated scientists, engineers, and journalists. She wore a funereal black A-line dress and blazer that matched her knee-high boots and the shoulder-length dark hair that framed her solemn face. A blood-red scarf encircled her neck. As always, she talked with a brisk, steely intensity that some of her colleagues found off-putting, though neither social disengagement nor a lack of compassion were its cause. Seager's mind seemed to be permanently overclocked, processing information faster and more keenly than most of her fellow human beings could fathom; her algorithmic approach to interaction, her abruptly earnest pronouncements, her calculated charm, all simply reflected that. She swept her eyes over the gathered crowd in the auditorium as she spoke, but often when offering her most fervent points, Seager paused to turn her piercing hazel gaze directly into the camera lenses, addressing an undefined audience of future generations.

She had brought the conference together, she said, to plot how to continue the field's wave of discovery in the face of the U.S. government's budget crisis and the seemingly dwindling exoplanet boom. "What we think here, most of us who work on exoplanets, is that hundreds or thousands of years from now, when people look back at our generation, they will remember us for being the first people who found the Earth-like worlds, and I don't mean Earth-size or Earth-mass. I mean Earth-*like*." Nearing her fortieth birthday, halfway through life, she said she no longer believed those discoveries to be foregone conclusions. "So I convened all of you here, and that's why we're recording this, because we want to make an impact and we want to make that happen. We are on the verge of being those people, not individually but collectively, who will be remembered for starting the entire future of other Earth-like worlds. That's why we're here."

It soon became clear that even if everyone agreed the field's sustainability depended upon searching for potentially habitable, potentially living planets around nearby stars, opinions strongly diverged about how such a search should take place. Charting a unified path through the coming years would be a struggle. David Charbonneau, a longtime friend of Seager's who was now a planet-hunting Harvard professor, rose from the crowd to make a case against pursuing a mission like TPF. Charbonneau had led the team that detected the first exoplanet atmosphere using Seager's technique. He wore a bright-yellow T-shirt emblazoned with the slogan "BIGGER THAN TrES-4," a reference to a transiting planet he had helped discover in 2007 that was so light and puffy it could float on water like a piece of balsa wood.

Oddball transiting worlds were one of Charbonneau's specialties; he had risen to prominence in 2000 when he codiscovered the first one, a hot Jupiter orbiting the Sun-like star HD 209458. Since 2009, he had spent much of his time on the mEarth Project (pronounced "mirth"), a ground-based array of small 0.4-meter telescopes that sought transiting super-Earths around nearby red dwarf stars, also known as M-dwarfs. The relatively large size of super-Earths in comparison to our own planet, paired with the relatively small size of an M-dwarf compared to our Sun, collectively meant that in terms of contrast the super-Earth/M-dwarf pairing was the easiest of all potentially habitable planetary systems to see, and would probably prove to be the cheapest kind to characterize. Charbonneau said that those transiting would be particularly good targets for transmission spectroscopy, as first outlined by Seager and others, all without needing to build anything like a multibillion-dollar TPF.

Such massive worlds would likely be quite alien, with unearthly

thick atmospheres and squashed landscapes due to their stronger gravitational fields. Unchecked by actual data, debates raged over whether
or not super-Earths could possess some form of climate-stabilizing
plate tectonics like our own more diminutive planet. To harbor liquid
water upon their surfaces, M-dwarf super-Earths would need to be perilously close to their small, dim stars, so close that tidal forces raised by
the nearby star would sap energy from the planets' rotations, causing
many to lock one face toward their stars just as the Moon does to Earth.
On such worlds, one light-bathed hemisphere would be eternally
scorched by ionizing radiation from stellar flares, while the other would
be veiled in endless night, with only a thin ribbon of constant middling
twilight between the two. Depending on its composition, a tidally
locked planet's atmosphere could entirely freeze out onto the night
side, or, if it persisted, drive steady gale-force winds between the disparate hot and cold hemispheres. If they were even habitable at all, no
M-dwarf super-Earth seemed likely to ever top a list of Earth-like exoplanetary real estate.

To Charbonneau, those environmental drawbacks and uncertainties were of little consequence, as was the fact that transit studies could
only reveal a vanishing fraction of the nearby population of exoplanets.
What was important was that transiting M-dwarf super-Earths could be
found and studied relatively soon, at low cost, without the need to wait a
generation or more. His argument was the distillation of a growing belief
among some in the exoplanet community that directly imaging Earth-
size planets in the habitable zones of Sun-like stars was a task so difficult
it was effectively a nonstarter. In place of TPF, a host of smaller, less ambitious, less capable ground- and space-based mission proposals had
emerged to sustain the astronomers during their travail in the budgetary
wilderness. Like Charbonneau's mEarth, most took their inspiration
from the wildly successful Kepler mission, and revolved around searching for transits around nearby stars. Two years later, NASA would allot
$200 million for the 2017 launch of one of those modest proposals, TESS,

the Transiting Exoplanet Survey Satellite. TESS would be the successor to NASA's Kepler mission, performing an all-sky search for transiting planets orbiting stars within a few hundred light-years of Earth.

Sharpening his presentation, Charbonneau pointed out that while there were only 20 Sun-like stars within about 30 light-years of our solar system, there were nearly 250 M-dwarfs. Extrapolating from Kepler results, which suggested that smaller, cooler stars harbored large numbers of close-in low-mass planets, Charbonneau stated that within at most 20 light-years of the Sun, "we are guaranteed that there are [potentially habitable] bodies at the right place around those M stars" to transit as viewed from Earth. Pushing for TPF was a mistake, he opined, not only because there was insufficient funding, but also because "it's foolish to devote twenty years of your life to something with too narrow a vision given the rate of discovery." In Charbonneau's view, younger astronomers would not and should not be willing to make such a lengthy investment in uncertain returns; missions like TPF and ATLAS would be doomed to wither and die on the vine for decades to come, and knowledge of any true Earth analogs would remain elusively out of reach. It couldn't be helped.

After a short break, Geoff Marcy, the doyen of American planet hunters, strode forward to tacitly critique Charbonneau's dismissal of big, challenging space telescopes, which he implied was misguided and counterproductive. He began his remarks with his hands shoved deep in his pockets, uncharacteristically gazing down at the floor as he restlessly shifted his weight from foot to foot. He was ecstatic about Kepler's results, he said, but angry about the last decade's lack of progress and the diminished prospects for the next. Kepler's results, he said, made the case for a TPF "extraordinarily compelling," for it suggested the existence around nearby stars of a wealth of non-transiting potentially habitable worlds that would otherwise elude close investigation. TPF-I in particular, with its promise of high resolutions obtainable only through unfilled apertures, was "the only plausible future for astrophysics," and yet "somehow NASA blinked." He spread his ire around the room, blaming not only NASA but also the vassal-like space-science community for profound failures of leadership. In the picture he presented, it was as if the agency and JPL had acted as interferometers, splitting the exoplanet researchers into competing, clashing fronts that canceled each other out in nulling pulses of self-annihilative incoherence. As a result, the collective dream of a TPF had been relegated to the dark fringes of space astronomy, and a deep shadow had fallen across the field's foreseeable future.

"This is the history as I know it painfully well," he recounted, recalling that he had served in 1999 on the inaugural TPF science definition team. "In 2000, NASA headquarters admonished us with a shaking finger from the administrator that we must build TPF-I, that we should all take classes, by the way, in astrobiology and molecular biology. . . . Then, around 2002, NASA said we should build a coronagraph, not an interferometer, and there's not money for both, so it *has* to be the coronagraph! And then, remarkably, in 2004 NASA headquarters announced that we should build both! A coronagraph in the optical and an interferometer in the infrared." He shook his head, incensed. "I didn't know how the money suddenly appeared to do both types of TPFs. We were being jerked around. For several years these two designs fought against each other, the coronagraph and the interferometer. . . . I think that was a pretty bad moment for a few years there. . . . And then of course the occulter came around, and the occulter cast a shadow, really, over the whole field!" The audience erupted in laughter, less from the simple joke and more to break the tension after the airing of so many uncomfortable truths.

Winding down his presentation, Marcy made it plain his faith in NASA had been so shaken by the failure of TPF that he now wondered whether portions of the agency's overlarge portfolio shouldn't be entirely outsourced to the more agile private sector. Besides TPF, he did dream of one other great and worthy task for the agency in the next half century, one he articulated in a Tsiolkovskian appeal directly to President Obama. "Stand up and make the following announcement,"

Marcy implored. "Say that before this century is out, we will launch a probe to Alpha Centauri, the triple star system, and return pictures of its planets, comets, and asteroids as soon as possible, even if it takes a few hundred years or a thousand years to get there. . . . It would be a great mission, to go to Alpha Centauri. It would engage the K-through-twelve children, it would engage every sector of our society, Congress, and so on. It would jolt NASA back to life, if we're really lucky. . . . And of course any such mission should be an international one involving Japan, China, India, and Europe. . . . A mission to Alpha Centauri would bring the diplomatic coherence in the world that we need, as well as scientific progress."

A few audience members laughed again, this time with bitter cynicism. Politically, the country was hyperpolarized, and financially it was mired knee-deep in debt. To think that any U.S. politician, let alone the president, would choose to expend one shred of political capital pursuing a voyage to the stars against such heavy headwinds smacked of the same wishful thinking that had years ago tipped the TPF project into calamity.

Moments later, Seager stood again before the crowd. She was scheduled to deliver a prepared talk but had discarded most of her presentation in light of the debate sparked by Charbonneau's and Marcy's remarks.

"We want to go out and map the very nearest stars," Seager reiterated, establishing the common ground on which everyone could agree. "Thousands of years from now, when people are embarking on their interstellar journey, they will look back and remember us as the people who found the planets like Earth around the very nearest stars. . . . I want to say I love NASA; NASA has helped my career tremendously. But I also see that NASA probably can't do the Terrestrial Planet Finder within the next forty years. That has become more and more clear to me, and everybody in this room knows that I am one of the biggest, ardent supporters of any terrestrial planet–type mission. That's what I want to do. I want a TPF in my lifetime. . . . And until now I never really worried about that." For a fraction of a second, her gaze and voice betrayed a sudden hint of sadness.

Seager noted that her position as a tenured MIT professor gave her tremendous security, which offered an opportunity—almost an obligation—to pursue high-risk, high-reward research. With flagging confidence that NASA could achieve TPF within her lifetime, she had been forced to consider other paths, new developments. One in particular looked promising: the recent debut of a new generation of commercial spaceflight providers, a select group of high-tech start-ups that were building rockets and spaceports with an eye toward at last overcoming the crippling paradigm of high launch costs. The companies had names such as SpaceX, Blue Origin, and XCOR, and multimillionaire CEOs who had made their fortunes with companies like PayPal, Amazon, and Intel. Seager thought the new companies might finally bring the profitable, sustainable human expansion into space that NASA had failed to deliver. They could be a powerful means to an elusive end, kicking off the next wave of synergies astronomers needed to lower the costs and accelerate the launches of TPF-style space telescopes, bringing the light of other living worlds into the lives and careers of all those assembled before her in the room. She had summoned her friends and colleagues to the meeting not only to discuss the field's future, but also to bid it a temporary farewell. Her work on exoplanets would continue, but it would compete against a new, overarching emphasis on aiding the emergence of a self-sustaining commercial spaceflight industry. The surprise announcement set off whispering reverberations through the crowd.

"That's what I'm going to be doing now," Seager explained with firm determination. "Most of you haven't seen me at meetings lately; you won't see me a lot in the future, because I'm investing in this. And if you see me working on asteroids and Mars, you'll know that I'm not really interested in those that much. I'm interested in getting the commercial spaceflight world whatever I can to help them." She cited estimates for launch costs: reaching orbit on one of NASA's space shuttles had cost some $100 million, while a ride on a simpler Russian Soyuz rocket was only $10 million. Commercial providers could perhaps drop launch costs by another order of magnitude. "We need them to succeed if we want to do Terrestrial Planet Finder, because we're never going to be able to do it at the ten-billion-dollar price tag. If we get that down by helping them, it will happen."

After Seager's talk, the audience filed out of the auditorium into the hallway, clumping into loose pockets of caffeinated conversation. I listened as a biochemist explained to an astrophysicist how the quest for a life-finding space telescope resembled the race during the 1990s to sequence the human genome. "There were all these different groups with the technologies to do it just trashing each other," said the biochemist. "Then you had the government agencies and the academic institutions and the pharmaceutical companies all separately decide to try to sequence it for their own ends. That mix of state and commercial competition pushed everyone toward the goal. . . . You guys need to figure out how to make China decide to go find the first habitable planets and name them all in Chinese."

Over by the coffee and tea tables, an engineer told a scientist that it would be straightforward to send a robotic probe at 10 percent of light-speed to Alpha Centauri: all he needed was a nuclear reactor from a Virginia-class submarine hooked up to a high-impulse electric propulsion system. "We could do it with today's technology!" the engineer exclaimed. "We'd probably be alive to see the pictures it sent back!" The scientist's only response was a polite nod, as if the engineer had forgotten to factor in a few important variables in his mental calculations.

That evening, after the conference's official end, a handful of the participants migrated from the Media Lab to Seager's office on the seventeenth floor of MIT's Building 54, the Green Building, the tallest high-rise in Cambridge. At Seager's invitation, some of us climbed up to the rooftop, which was dotted with antennae and white radar domes, to gaze down at the twinkling lights of Boston's skyline and at sailboats plying the calm waters of the Charles River. Mountain, Seager, and

Grunsfeld chatted quietly, admiring the view. Traub stood silent for a time, watching the sunset. Marcy clambered up to pose for a few pictures beneath the immense radar dome, then descended. He mostly made small talk, but, when pressed, would dive again into discussing NASA's plight.

"NASA's in big trouble," he told me later. "It seems like even with all its infrastructure and expertise, it can't outdo the private sector. It's unable to overcome its own bureaucracy. How could NASA turn its back on TPF? I don't want to blame NASA per se; maybe it's not really NASA's fault. Maybe it's just that we have challenges when we try to organize ourselves to do great things. Rome falls. People are imperfect. We make incredibly tragic mistakes. . . . It's just our nature, it seems." He raised an empty hand, with thumb and forefinger angled like chopsticks grasping a single grain of rice. "We are just this far above the ants. That's how I see it. We function in some ways like a bee colony. It's natural. But, you know, there is something called colony collapse disorder, too."

Back inside the Green Building, the discussions continued, and a drift of people buzzed around Seager, the queen bee of this transitory hive. Standing at its fringes, I overheard her muse again about interstellar travel. "I don't know if we'll ever leave the solar system," she said. "All I know is, it would be nice to have the option."

CHAPTER 10

Into the Barren Lands

An expanding shell of light surrounds our solar system, with our Sun as its source. The shell is not perfectly spherical, but instead tapered like an hourglass at its midsection, where some light is extinguished by thick lanes of gas and dust within the Milky Way's spiraling galactic plane. Above and below the galactic plane, relatively free of occluding debris, the Sun's photonic shell ripples outward in twin lobes, ever expanding at the speed of light. Though the shell's boundaries sweep three hundred thousand kilometers farther away from us every second, their expansion through the great intergalactic voids is so glacial that their position can be pegged at 4.6 billion light-years away. The shell's edges are composed of photons first erupted in the flash of thermonuclear ignition that announced our star's birth. Each unfolding moment of our solar system's history follows behind, encoded in planetary reflections, refractions, and occultations of starlight. In all probability, the beginning of the end for this photonic broadcast will occur some six billion years from now, when our Sun, long since swollen into a pulsating red giant, finally burns through its last stores of hydrogen and helium. It will leave behind scorched planets, an evanescent nebula of ionized gas, and a stellar remnant, a white-hot ember of carbon and oxygen ash. Slowly cooling over the eons, the remnant's faint light will finally fade to black, switching off the solar transmission as surely as scissors cutting a thread, leaving only the light of ancient days to echo through eternity.

Borne on photons, the echoes of primordial and Precambrian time—the formation of planets, the emergence of life on Earth, the oxygenation of our world's atmosphere, the invasion of the land—all long ago left the Milky Way to wash over the surrounding galaxies, galactic clusters, and superclusters. An observer somewhere among the trillion stars of Andromeda, our nearest neighboring spiral, would today see the Earth of 2.5 million years ago, when the forerunners of *Homo sapiens* were perfecting the production of crude stone tools in sub-Saharan Africa. Seen from the Large Magellanic Cloud, a dwarf galaxy swooping near the Milky Way, our world would be locked in the glacial advance of 160,000 B.C., with our ancestors poised to migrate out of Africa as the ice sheets retreated. Within our own galaxy, the echoes are closer to home. Among the open clusters and blue hypergiant stars of the Carina Nebula, somewhere between 6,500 and 10,000 light-years away, the Earth appears as it was during the rise of agriculture and the Bronze Age civilizations of Mesopotamia, Egypt, and the Indus Valley. Light from the Earth of Thales, Democritus, and other ancient Greeks now washes over the blazing newborn stars and shimmering molecular clouds of the Christmas Tree Cluster, just over 2,500 light-years distant. The Earth in the skies of the giant planets circling the Sun-like star HR 8799 has just begun transmitting in radio and perfecting the internal combustion engine. The first television transmissions of the 1930s now roll over the ice-blue stars of Regulus, and news of 1969's *Apollo 11* lunar landing has just reached the aging yellow suns of Capella. Whether any of this has actually found an audience somewhere out there, we cannot yet say. For all we know, the lively broadcast from Earth may be the only one of its kind in the observable universe.

Viewed from the vicinity of the closest stars and compressed into a short time-lapse movie, our solar system's birth and evolution would present an eerie picture. From a large black cloud of molecular hydrogen, a star forms first, followed by whirling planets. Once settled in their orbits, the outer giant planets remain relatively inert, placid for billions of years beneath their whorls and bands of swirling gas. Even less happens on inmost Mercury after its magma oceans cool and crust over. The other three inner worlds are each a blue-green jewel of cloud, sea, and land, but in a flash Venus bakes beneath a pall of steam, and Mars withers and freezes. For most of the movie's running time, Earth is the system's most curiously variable world, a kaleidoscope of wandering continents, pulsing glaciers, erupting mountains, surging tides, and swarming greenery. In the last second before the time-lapse catches up to the present day, the Earth gains electric lattices of nocturnal lights and sparkling haloes of artificial satellites. The transformed planet ejects a handful of spore-like metallic flecks throughout the system. Five of them approach Jupiter and are flung off at solar-escape velocities, destined for parts unknown in the wider galaxy and cosmos. They are humanity's fledgling interstellar probes, each launched by NASA: *Pioneer 10* and *11*, *Voyager 1* and *2*, and the Pluto-bound *New Horizons*.

On February 14, 1990, the farthest and fastest of those probes, *Voyager 1*, turned its cameras back toward Earth for a final time from a distance of more than six billion kilometers, beyond the orbit of Pluto and high above the solar system's ecliptic plane. At the insistence of Carl Sagan and other workers on the Voyager missions, the spacecraft sought to reproduce the iconic Apollo "Blue Marble" image, but from a distance one hundred thousand times greater. From so far away, the entire

Earth was almost lost in the Sun's diffractive radiance, but close inspection revealed our planet as a solitary azure point of light comprising less than a single pixel in *Voyager 1*'s transmitted image.

Sagan called Earth's image a "pale blue dot," and went on to use the phrase as the title for one of his many bestselling books. In the decades since the Green Bank meeting, he had ascended to the pinnacle of practicing and popularizing space science, performing crucial work on planetary atmospheres and producing the wildly successful television miniseries *Cosmos*. With Frank Drake and other collaborators, Sagan had designed and curated a long-playing phonograph record to be sent to the stars with the Voyagers. Crafted from copper, aluminum, and gold, a copy of the record was bolted to the side of each spacecraft, ready for playback, complete with a magnetic cartridge, stylus, and pictogram instructions. In the emptiness of interstellar space each record should persist for untold eons, outlasting the Sun and the Earth alike. Any eventual encounter with another planetary system is improbable—if ever found at all, the records will most likely be recovered by some wildly advanced civilization traveling between the stars. Maybe even by our distant descendants, if we are so lucky. The Voyager records were vaingloriously utopian, and excluded references to such entropic human failings as crime, war, famine, disease, and death. Each contained recorded messages from President Jimmy Carter and United Nations diplomats, greetings in fifty-four languages, and 118 joyful photographs of life on Earth. Each would share the sounds of wind and rain, heartbeats and laughter, kisses and rocket launches, electroencephalograms and whale songs. Each would play the music of Bach and Beethoven, Mozart and Stravinsky, Peruvian panpipes, Javanese gamelans, and Chuck Berry performing "Johnny B. Goode" on his electric guitar. Each would be a murmur from the departed Earth, a golden memory of beings either sublimed into some unknowable future form or long fallen from their ancient flaws.

To many humbled and earthbound souls living in a universe revealed year by year as increasingly aloof and uncaring, Voyager's glimpse of far-off Earth and its messages to the stars became beacons of hope, perseverance, and wisdom, pure and noble expressions of the better angels of our nature. Meditating on the pale blue dot in an essay, Sagan poetically called it a "mote of dust suspended in a sunbeam." It was "the aggregate of our joy and suffering" upon which "everyone you love, everyone you know, everyone you ever heard of, every human being who ever was, lived out their lives." To Sagan, the image was a symbol of the cosmic folly of human divisions and geopolitical conflicts. "In our obscurity, in all this vastness, there is no hint that help will come from elsewhere to save us from ourselves," he wrote. "The Earth is the only world known so far to harbor life. There is nowhere else, at least in the near future, to which our species could migrate. . . . [T]he Earth is where we make our stand."

Earlier in the same essay, Sagan touched on the difficulties of finding any plausible future planetary home for humanity. He thought that the pale blue dot approximated the view of Earth as seen from a starship arriving after a long interstellar voyage. He did not mention that it also replicated how our planet would appear through a first-generation TPF-style space telescope, though the thought perhaps crossed his mind. We would know from experience that our home planet's pale blue came from its life-giving seas of water and clouds of water vapor, Sagan wrote, but he doubted whether an alien observer could surmise so much from Voyager's single, spectrum-free image. More detailed inspection would be required.

That inspection arrived ten months after *Voyager 1*'s historic image—on December 8, 1990, when Sagan masterminded a suite of Earth observations using the Galileo spacecraft, which was flying by our planet on a roundabout voyage to Jupiter. Examining the Earth as if it were a newly discovered alien world, Sagan and the Galileo team successfully confirmed Earth's habitability, then detected its biosphere and technology, all from the depths of space and purely from first principles. They took the planet's temperature in infrared light, and confirmed its polar caps, seas, and clouds were made of water.

They found evidence of life in the oxygen-soaked, methane-tinged atmosphere, far out of thermodynamic equilibrium, as well as in the vegetation-filled continents, which reflected the spectral sign of light-absorbing chlorophyll, of photosynthesis, out into space. Powerful pulses of narrowband modulated radio waves from the planet's surface hinted at technological civilization. The collective verdict was indisputable: much of the planet was literally covered with life, and something down there had been smart enough to build a global telecommunications network. Later, Sagan and his team turned Galileo's instruments toward Earth's Moon, finding to no surprise that, unlike our living planet, it was a desolate, dead rock. As pat and prosaic as Sagan's Galileo observations may first appear, they constituted a potent control experiment, a standard of proof that could be applied to any planet, whether investigated via close-flying probe or by telescopes gathering light over interstellar distances.

Examining the breadth of Sagan's later work, it is hard to escape the conclusion that he was methodically preparing for any observational studies of potentially habitable exoplanets that might occur during his lifetime. We will never know for certain. His life was cut short in December 1996 at the age of sixty-two, after a two-year battle with bone marrow disease, only months after NASA's administrator, Dan Goldin, had announced plans to build TPF. Even at the end, by all accounts Sagan was just as sharp and limber-minded as he had been in all the earlier decades of his scientific career. If Goldin's initial projections for TPF's launch in 2006 had held, Sagan would have been seventy-two when the telescope began uncovering any pale blue dots around nearby stars. Had he lived longer, he could have served as an authoritative elder statesman to guide and promote the next giant leap in humanity's understanding of the universe. Instead, with Sagan's passing and the eventual demotion of TPF to NASA's technology-development dustbin, his Voyager and Galileo observations of Earth were quite possibly the closest astronomers would get to investigating a living, alien world for many generations to come.

• • •

In 1990, while Sagan was scrutinizing Earth from afar, Sara Seager was beginning her freshman year at the University of Toronto, tearing through her introductory coursework in math and science. She had convinced her father, a doctor who had left medicine to start a small hair-transplant business, that she would pursue a premed track. He encouraged her to specialize in something lucrative, dependable, and relatively stress-free, like dermatology. Instead, to her father's chagrin, Seager soon switched her focus to physics and astronomy. She had been curious about the night sky ever since she was a young girl, when on nighttime family car rides she would wonder why the Moon always seemed to follow overhead no matter where they went. Soon after, Seager's father took her to a "star party," where an amateur astronomer explained the Moon's orbit and let her gaze at it through a telescope. When she was ten, on a camping trip into the Canadian backwoods, Seager's view of the world had drastically expanded as she stepped out of her tent at night beneath a clear sky suddenly free of city lights. Looking up at so many stars, she for the first time sensed the continuum that began with the Earth beneath her feet and extended out into the endless heavenly depths above. At sixteen, while attending a university open house, she learned that some privileged people actually studied stars, planets, and all else beyond Earth for a living.

"It was one of the most exciting days of my entire life," Seager later recalled to me. "You can do this as a job? I rushed home and told my dad. He was so hard on me, and discouraged me with the harshest lecture he ever gave. He said, 'You have these natural skills, but you need to be able to support yourself and not rely on any man!' He wanted me to be independent, and just didn't think it was a good career choice." Seager's father valued practicality, but time and time again, he told her she must think big, set goals, and visualize herself reaching them. Otherwise, she should not expect success.

Despite that advice, Seager often described her early path toward astronomy as an unfocused "random walk," like that of a photon bouncing chaotically around the seething heart of a star. She appeased her father by first concentrating on physics, reasoning that would boost her chances of employment both within and outside of academia, but the more she learned, the less interest she could muster. "I believed you could perfectly describe everything with equations," she said. "Then I learned that approximations were rampant. I was working so hard for three, four years, why should I suffer my whole life and work so hard for something that's not enjoyable?"

Approaching graduation, she took a risk and tried her luck applying to astronomy programs at graduate schools. She decided to think big, and submitted an application to Harvard in the fall of 1994. She was twenty-two. To her astonishment, Harvard replied in February of 1995, offering her a grad-school spot as well as a modicum of funding. Seager received the news while cross-country skiing with friends in Ontario. She accepted, and set about planning her move from Toronto to matriculate at Harvard in the fall. That summer, she had little to do other than wait. She decided to travel north and go camping, but she didn't want to journey alone. Seager reached out to an occasional canoeing partner, Mike Wevrick, a robust thirty-year-old who loved cars and the outdoors. Wevrick looked rather like a grizzled Marine, blue-eyed and broad-shouldered, with long, powerful legs and biceps as big as Seager's thighs. He sported a crew cut and a constant few days' growth of stubble on his lean face, and had a reputation for quiet intelligence and kindness. They had first met skiing in Ontario, on the very day Seager learned of her Harvard acceptance. For both reasons, she would later call it the luckiest day of her life.

Together, Wevrick and Seager devised an ambitious canoeing trip deep into Canada's Northwest Territories, to the "Barren Lands," the treacherous tundra that exists past the northern limit where trees can grow, a trackless wilderness so desolate and remote that it was essentially unmapped until after World War II. They would begin by driving four and a half days from Toronto into the boreal forest of northern Saskatchewan, to where the northerly road ended at a lake. From there, for twenty days they would canoe farther north on a series of rivers before finally reaching Kasba Lake Lodge, an outpost with a small airstrip where they could replenish their supplies. From Kasba Lake, they would continue canoeing north past the tree line, into the Barren Lands, on an out-and-back journey that would take another forty days. They planned to be back at Kasba by the end of those forty days to catch a plane south. Wevrick was an expert whitewater paddler, and would guide them through the rivers and lakes. Seager would help with portages, the overland hauling of supplies and Wevrick's red Old Town Tripper pack canoe between navigable waterways. They left Toronto on June 24, at the end of the summer thaw, planning to return in late August as northern autumn fell.

The weeks leading up to the trip had been arid, with hardly any rain. Departing Toronto, they felt optimistic the weather would hold, minimizing muddy slogs and soggy supplies. But the lack of rain also raised the risk of lightning-sparked forest and prairie fires. Arriving at the end of the road, they found the lake and surrounding forest blanketed in dense smoky haze. They paddled into the gloom and through the mouth of a river, stopping to tie wet T-shirts around their faces when they passed smoldering shorelines. They developed a routine, taking meals in the canoe and paddling through most of the twenty hours of daylight provided by the sub-Arctic summer Sun. If the wind blew at their backs, they rested and used a plastic tarp as a makeshift sail. They portaged as often as fifteen times a day to circumvent sequences of boulder-filled rapids and plunging waterfalls. When they ventured onto land, swarms of biting black flies and mosquitoes rose from the underbrush to assail them. When daylight faded, they made camp and fell into weary, dreamless sleep within their tent.

Seager and Wevrick found comfort in their quiet conversations, and in the harsh, untrammeled beauty of the surrounding wilderness. They walked upon the Precambrian, Archean, Proterozoic rock of the Canadian Shield, the oldest exposed rock on planet Earth. They portaged over the roots of tall mountains transformed to gentle nubs by four billion years of weathering, in a country that had been compressed and scraped clean by the weight of advancing ice sheets during the last glacial advance. Sediment-clogged streams and rivers of meltwater had run like veins beneath the ice sheets, so that when the ice withdrew, it left behind eskers—sandy ridges of pink granitic gravel that followed the twisting paths of the dried-up subglacial flows. The eskers wound between and around kettle lakes, each lake a ghostly puddle from some great hunk of ice long ago calved off the retreating glaciers. The land was still slowly rising up and finding its bearings, rebounding at one centimeter per year from the heavy load of ice that had been lifted tens of millennia ago. Distant columns of smoke from incessant fires lined the horizon in all directions. During their twenty-day trek to Kasba, they saw abundant wildlife, but not another human being outside of each other.

In mid-July, Seager and Wevrick reached the Kasba Lake Lodge, perched on the west end of an island-dotted expanse of water stretching as far as they could see. They picked up supplies, hobnobbed with the lodge's caretakers, then continued north, up through the lake and into the Barren Lands. Day by day, mile by mile, the trees became sparse, then stunted, then entirely absent, replaced by moss carpets, hardy grasses, and brightly colored lichens. Just north of the tree line, they spotted their first caribou, gazing at them as if they were visitors from another planet. Without trees, a constant wind rushed unrestrained over the sinuous hills and through the rivers and lakes in the hollows. The wind made canoeing slow going, often pressing them ashore at midday. Sometimes on a windy shore Seager would pull an astrophysics textbook from her pack. Far below the tree line, in another world, Harvard awaited. Other times, she and Wevrick would have long talks, seemingly the only people in existence, inhabiting a universe built for two.

The sky overhead was hazy when they reached the northernmost point of their journey. Even though they were more than 200 kilometers north of the tree line, smoke from the fire-ravaged southern forests still reached them on strange, unseasonal winds. Atop a treeless knoll, they stumbled upon five stone cairns. Old Inuit graves—the first sign of humans Seager and Wevrick had seen since Kasba. Scavengers or looters had scattered some of the rocks aside, revealing artifacts of rusted metal and wood, as well as a small sun-bleached human skull. Seager snapped a photograph. She wondered what the person had looked like, how they had died, and why they had chosen to live so far from the world she knew. She looked up from the skull to the surrounding hills, spotted with pale grass and summer wildflowers, rolling on and on. The silence was broken only by the whisper of wind rippling from horizon to horizon. Silver and blue circles of sky pooled in the clear, cold water of countless lakes. In that moment she understood why someone might abide in such everlasting solitude.

After crossing back south beneath the tree line, they entered what Seager called "Esker Territory," a seemingly interminable warren of sandy pink hills coiled in complex double and triple ridge systems, with small lakes and woods lying in the valleys between. It was beautiful, but exhausting to cross. The days blurred, and the geography rolled by like even-spaced rumble strips on a lonely highway. All was esker. Then lake. Forest. Esker. Lake. Boulder-filled streambed, esker, and lake. They smoothly, silently paddled and portaged for hours, now adapted to the rhythm of the land long ago laid down in the ebb and flow of ancient ice. No words were needed. They were in unison, almost reading each other's thoughts. On one of their last evenings in the far north, Seager stood alone atop a spruce-sheltered ridge, contemplating the blue lakes and pink eskers as the Sun sunk lower in the sky. They were in a different world, one made all the more real by its distance from the bright, baleful cities and ever-scurrying crowds. Perhaps someday the cities and crowds would encroach here, too, driven poleward by drowned coasts, but for now the land lay empty. They had seen no one else for over a month, yet they were not lonely. They ate when hungry, slept when sleepy, and lived simply, yet never yearned for more. "We had grown so content with each other's company that we had no psychological or emotional cravings for anything or anyone from 'outside,'" she later wrote. "The trip became our perfect life."

When their plane lifted off from the airstrip at Kasba on August 28, Seager looked down with wistful thoughts at the windswept lake and the little rivers that blindly meandered through the grasslands and conifer woods. "In sixty days, 'real' life had become so dim as to seem partly impossible and mostly unbearable," she would recall. "The solitude, the vast wilderness, the free and compelling lifestyle, the constantly changing terrain, and my excellent companion were a truly unbeatable combination." She realized she had not only fallen in love with remote desolation; she had fallen in love with Wevrick, too. Soon after they returned, Seager asked him to move with her to live together in Cambridge. Without hesitation he agreed.

At Harvard, Seager initially focused on cosmology, specifically the basic physics behind "recombination," an event that occurred less than a million years after the Big Bang. Back then, our universe was still just a hot expanding mass of plasma, an opaque fog of electrons and protons with no atoms, no molecules, no stars and galaxies. For hundreds of thousands of years the plasma cooled and expanded, until it reached a critical transition, becoming cold enough for electrons to "recombine" with nuclei, glomming together to form atoms. In a literal flash, the atoms froze out of the primordial plasma, transforming the expanding fog of plasma into a transparent cloud of hydrogen and helium, unleashing a flood of light that still reverberates through the universe today. We detect it as an omnidirectional all-sky glow of microwave radiation with a temperature less than three degrees above absolute zero. As Seager worked on recombination, the first discoveries of hot Jupiters were trickling in. She approached her advisor, Dimitar Sasselov, looking for ways to segue into exoplanets, which she saw as a more interesting topic. Sasselov steered her toward modeling hot-Jupiter atmospheres, since, as with the epoch of recombination, the associated calculations partially concerned the mechanics of high-temperature hydrogen and helium. From that seed sprang Seager's subsequent PhD and her career-defining early work that led to the first detection of an exoplanet's atmosphere.

Meanwhile, Wevrick forged a successful career of his own writing and editing high school science and math textbooks. Throughout Seager's Harvard tenure they escaped the city for the countryside whenever they could, and they eventually married in 1998, the same year that Seager completed her PhD thesis. The following year they relocated to Princeton, New Jersey, where Seager had secured a five-year fellowship at the Institute for Advanced Study, the same establishment where Einstein had spent the last years of his life. There, with the encouragement of another mentor, the late astrophysicist John Bahcall, Seager began meeting with several exoplanet-oriented researchers at nearby Princeton University, developing concepts and techniques that could be used to characterize exoplanet atmospheres and surfaces with one of NASA's forthcoming TPF telescopes.

After one such meeting, Princeton's David Spergel was inspired to conceive the coronagraphic masks that became a technological pillar for TPF-C. After another, Seager and two Princeton astronomers, Eric Ford and Edwin Turner, devised an exquisite method to gain information about an Earth-like exoplanet solely from the fluctuating brightness of its pale blue dot as seen across interstellar distances. They began by developing a model to calculate the amount of scattered starlight any given planet could project toward a distant observer, and as a test case ran it based on Earth-observing satellite data. As our virtual pale blue dot turned in various viewing geometries beneath their model's scrutiny, over time the team found they could discern what region of the planet they were looking at solely from its brightness, despite its reduction to an unresolved starlike point.

Looking down on the equator, for instance, each day like clockwork the relatively bright continents of North and South America would rotate into view, sandwiched on either side by long dark stretches of open Atlantic and Pacific Ocean. In their repetition, such patterns revealed the length of Earth's days. With the rotation rate established, Seager, Ford, and Turner could attempt more granular mapping, trying to discern the bulk fraction of ocean versus land, as well as finer features such as forests, prairies, deserts, and ice sheets. They feared bright reflective clouds would confuse their observations, but they found that clouds tended to arise and dissipate in predictable ways—more often at land-sea interfaces and less frequently over open ocean and dry continental interiors. They learned to distinguish the reliably cloud-free Sahara Desert by the intense near-infrared brightness of its hot sand and the lush, verdant Amazon Basin by its constant blanket of clouds. They saw hints of ice sheets, lakes, and seas by occasional spikes in brightness, when their smooth, flat reflective surfaces glinted sunlight back into space like mirrors. Given enough time, they suspected, they could even discern the varying reflectances of shifting vegetation, clouds, and ice cover that would come through changes in weather, seasons, and climate. All purely from a single wavering point of light, without the need to first obtain planetary spectra using an 8- to-16-meter mirror in space. Of course, they had the advantage of already knowing what they were looking at; teasing apart such features for the unknown environment of an actual faraway terrestrial exoplanet would be much more difficult. But the technique offered hope that even a relatively small 2- to-4-meter space telescope might be able to roughly map any Earth-like planets around the handful of closest stars. Seager pressed on, churning out a series of papers outlining how extremely precise measurements of transits could reveal properties such as an exoplanet's rotation and atmospheric structure.

Now midway through her fellowship, Seager began searching for what would come next. Despite her leadership in the rapidly growing field of exoplanets, she received polite dismissals from many potential employers, who seemed to believe Seager's optimistic visions of finding other Earth-like worlds would never come to pass. The exception proved to be the Carnegie Institution, which offered her a job in 2002. With Bahcall's blessing, she left the Institute for Advanced Study and moved with Wevrick to Washington, DC. At Carnegie, she became even more involved with planning for NASA's TPFs, and was for the first time exposed to the rigor of geophysics. Seager began exploring how to theoretically and observationally constrain not only an exoplanet's surface and atmosphere but also its deep interior—things like its bulk composition, or its likelihood of volcanic activity and plate tectonics. Transits were key, since they allowed astronomers to measure a planet's radius, its size. Paired with mass estimates from radial-velocity measurements, this yielded a planet's density. Seager and others developed mass-radius relationships for worlds of various compositions, estimating how planet hunters could distinguish between, say, one Earth-size planet made of pure water and another composed of mostly carbon, or iron. The work would later prove crucial as more and more worlds of intermediate dimensions were detected. When they transited and their densities were calculated, many of the so-called super-Earths astronomers were finding proved in fact to be "mini-Neptunes," gassy worlds with thick, opaque atmospheres of hydrogen and steam, rather than rocky planets with thin layers of translucent air.

As an emerging leader in the burgeoning field of exoplanetology, Seager began receiving frequent invitations to speak at high-profile conferences, meetings, and colloquia, and her retreats to the wilderness with Wevrick grew few and far between. In 2003, the trips for work and play were both sharply reduced—Seager became pregnant, and gave birth to their first child, a boy. They named him Max. Another boy, Alex, followed two years later.

By the autumn of 2006, though the field's fortunes had fallen when NASA pulled the plug on TPF, Seager's star was continuing its rise. MIT had lured her from Carnegie with an offer of immediate full tenured professorship—the equivalent of a lifelong golden ticket for any academic, but one particularly valuable for a researcher so young and just starting a family. Seager and Wevrick placed a mortgage on a grand old house in Concord, Massachusetts, a fixer-upper not too far from Walden Pond. She would begin her professorship in January of the New Year. Seager was pleased with her progress, and broke the news to her father on a visit back home. He had recently been diagnosed with terminal cancer, and though he was fighting hard, they both knew he was in rapid decline. Her gamble on astronomy had worked out, she said. She was thirty-five, already with tenure at one of the world's premier institutions—she told him it was the best she could expect to do. Seager hoped he would be proud. Instead, he transfixed her with an icy stare and answered slowly, with a voice like cold steel, "I never want to hear you say that anything is the 'best' you can do," he said. "I never want you to be limited by your own negative thinking. I know there's an even better job, and I know you'll get that one too, someday." Not long after their talk, her father died. To the very end, he pushed Seager to never stop thinking big.

At MIT, she began thinking bigger than ever before, assembling several research groups and pursuing multiple different initiatives designed to extend her expertise from theory into observation, engineering, and project management. To have any hope of being at the helm of a future TPF, she would need experience in all four realms. Personally, professionally, her focus was fixed on the future—every day, it seemed, the boys grew, looking more and more to her eyes like their father. Wevrick taught Max and Alex how to paddle a canoe, bait a hook, and make a fire. Seager taught them, too. She would tell her wide-eyed boys about the origin of the Sun and the Moon, the history of the Earth and its companion planets, and the newly discovered worlds that circled so many stars like grains of sand. Max loved logic and numbers— perhaps he would become a mathematician. Alex enjoyed puzzles and games, and like his parents was drawn to the outdoors. Perhaps he would be an artist, an inventor, or a forester. By the time they were men, she thought, NASA could again be preparing for TPF. She would be ready, having raised a family and acquired new skills in the interim. Her life, intertwined with Wevrick's, was coming into fuller bloom than she could have ever hoped or planned.

In late September of 2009, Wevrick began to notice a dull pain and occasional sharp cramps in his lower abdomen. It seemed to flare up randomly—he could find no correlation between the pain and anything he did. At first he didn't worry much—after all, he regularly exercised, ate healthy food, and didn't smoke—but after weeks of discomfort he began consulting medical websites, yielding indeterminate results. By mid-November, the pain had grown worse, and he was worried enough to seek advice from friends. His friends suggested one malady after another: appendicitis, inflamed gallbladder, irritable bowel syndrome, ulcers, diverticulitis, a hernia, Crohn's disease. None perfectly matched his symptoms, which were stubbornly general. Seager convinced him to see a doctor, who, after cursory poking and prodding, found no signs of serious illness. Over the next two months, he experienced a few bouts of pain and vomiting, which he assumed to be food poisoning. In mid-January of 2010, he suffered another attack, more severe than before, and ended up in the emergency room. A CAT scan, colonoscopy, and biopsy revealed grim news: a large mass of what appeared to be cancerous cells had blocked most of his small intestine. He'd had Crohn's disease after all, asymptomatic and undiagnosed for years, but the chronic inflammation had finally sparked cancer.

Surgery in early February excised the growth and surrounding tissue, and Wevrick began successive rounds of aggressive chemotherapy. He took it all in stride, displaying the same conservative equanimity he had relied on in the past when he faced life-threatening conditions in the wilderness far from home. He even felt well enough to go whitewater canoeing in Idaho in July. But by October, the cancer had returned and metastasized with renewed virulence, resuming rapid growth. Seager busied herself scouring the medical literature and consulting some of the nation's top experts in small-bowel cancer. Perhaps there were new experimental treatments to be tried outside of the United

States; maybe they could move to Europe to take part in some long-shot clinical trial. The experts gently downplayed such thinking; there was little hope. Seager told Wevrick they could drop everything, travel the world, do whatever he wanted while they still had the chance. Just say the word, and they would go. For now, he replied, he didn't want to be disruptive, and was comfortable at home. He thought there was still time.

Meanwhile, Seager still needed to work; she could not allow herself to crumble into grief. She made arrangements for babysitters, and found nurses to provide palliative care. Having watched her father succumb to cancer, she knew this was the calm before the storm. Some evenings she would walk to nearby Walden Pond, to the same still water and sweet scents of oak and hickory that the Transcendentalists Ralph Waldo Emerson and Henry Thoreau had so cherished more than a century before. One day, she promised herself, whether with her two boys or any grandchildren, she would stand beneath the dark sky of Walden Pond and, pointing to a bright point of light, tell them that star possessed a planet very much like the Earth. "Each time you look up at it," she would say, "someone there may be looking right back." The thought gave her solace, and a feeling of being very big and oh so small, all at once. She would not give up; she would endure and grow stronger. She would help map the nearby stars; she would seek out those other Earths. In those moments, the death and loss that surrounded her would shrink, becoming infinitesimal in a vista of such magnitude its full expanse surpassed all sight.

"Men frequently say to me, 'I should think you would feel lonesome down there, and want to be nearer to folks, rainy and snowy days and nights especially,'" Thoreau wrote in *Walden*, his classic 1854 chronicle of two years spent alone on the pond's shores. "I am tempted to reply to such—This whole Earth which we inhabit is but a point in space. How far apart, think you, dwell the two most distant inhabitants of yonder star, the breadth of whose disk cannot be appreciated by our instruments? Why should I feel lonely? is not our planet in the Milky Way?

This which you put seems to me not to be the most important question. What sort of space is that which separates a man from his fellows and makes him solitary?"

By March of 2011, Wevrick sensed time was running out. He drew up an orderly three-page list of practicalities: upkeep tips for the house and the car, contact information for relatives and their life insurance agent. Grim milestones ticked by, like life's beginnings glimpsed in a rearview mirror: the last days of walking, sitting up, speaking, moving. He fought each step of the way with characteristic strength, but death would not wait. The home nurses brought in more medical equipment, with long tubes and beeping monitors, as well as a hospital bed. A final vigil began only days before Seager's planned "Next 40 Years of Exoplanets" conference, where she announced her strategic career shifts in pursuit of TPF. At night, she recalled, she would curl beside Wevrick in his bed to talk as he drifted in and out of fog. She whispered that she loved him, that he had changed her life forever for the better, that everything would be all right, that he could let go. She told him he had inspired her to take the risks required to achieve her dream and change the world. Wevrick cracked a wan smile. "No," he said, shaking his head. "You would have done that anyway." It would be one of their last conversations.

At their home, two days after Seager's fortieth birthday, Mike Wevrick passed on into peace, Seager by his side. Their long journey together had come to an end.

Seven months later, I was in Seager's seventeenth-floor office overlooking the Charles River. She sat across from me in an overstuffed red chair, radiant in a beam of bright morning sunlight streaming through the large windows. Behind her stretched a floor-to-ceiling, wall-to-wall chalkboard, covered with arcane notations and diagrams. Seager was in the midst of an ambitious new project: attempting to quantify the alternate biosignatures that could manifest on the wide variety of unearthly habitable planets that might exist. She looked good, relaxed, with an easy smile. I told her so.

"Thanks," she said, her smile fading. "I feel terrible. A little depressed."

The month after Wevrick's death had been a blur, Seager said. She had found a small support group of widows in Concord who occasionally met to socialize and share their stories. She had planned a series of trips to spend reconstructive time with Max and Alex—they watched NASA rocket launches in Florida, hiked in New Hampshire and Hawaii, camped out in the Southwest, visited the Smithsonian in Washington, DC, and traveled through Europe. She had also poured herself into work with renewed vigor—she felt there was little other choice.

Other than her children, TPF remained Seager's dream, goal, and driving force. "There are three paths to it," she said, looking at me intently. "One is NASA, the government way, where I'll put myself in the right position so that I could be a principal investigator in the future. The second way is to pursue it through some private initiative. And the third way is, I'll simply make so much money that I can literally fund it all by myself."

She had methodically devised overlapping preparations for each path, and explained them to me in more detail. An MIT/Draper Laboratory project she helmed called "ExoplanetSat" lay at the crux of Seager's interlinked plans. Already well along in development, ExoplanetSat was a "nanosatellite," a golden metal rectangular box no bigger than a loaf of bread, jam-packed with a tiny telescope, deployable solar panels, and a miniaturized avionics package for precision pointing and ground communications. It was designed solely for the purpose of constantly monitoring a single nearby Sun-like star for any signs of transiting planets, and possessed sufficient sensitivity to detect planets just a bit larger than Earth. The first ExoplanetSat would cost some $5 million to develop and launch, but subsequent copies could practically roll off the assembly line at a cost of a half million dollars apiece—dirt cheap for any hardware destined for orbit. Each would operate in low Earth orbit for a minimum of one to two years. Nanosatellites are so small that they typically do not require dedicated launch vehicles; instead, they piggyback on rockets launching larger primary payloads. Seager envisioned eventually launching a whole fleet of low-cost ExoplanetSats to surveil all the nearest, brightest stars for potentially habitable transiting planets. The first prototype was set to launch as early as 2013 as part of a NASA program supporting nanosatellites.

The practical expertise she would gain in engineering and management from ExoplanetSat's success would make Seager a more appealing choice for involvement in future NASA missions, but would also serve as a stepping-stone for her own development of more ambitious spacecraft. Her second path, the private route, involved raising money to build and launch a downsized and simplified TPF that could survey the nearest hundred Sun-like stars for any exoplanets. Such a telescope would not be large or sophisticated enough to gather spectra for any habitable worlds, Seager said, though it could potentially characterize them via the photometric techniques she had pioneered earlier in her career. For the second path, she had already found a powerful partner, Scot Galliher, a fifty-something technologist who decades earlier had cofounded Goldman Sachs's Financial Institutions Technology Group. Together they had recently formed the nonprofit Nexterra Foundation to pursue a private planet-finding space telescope, though they were still sorting out its finer details.

"Nexterra's goal is to map the nearest Sun-like stars, no more, no less," Seager told me. "Maybe it only finds the pale blue dots, and then the next generations will get their spectra, maybe even find a way to go there. It's unconventional, but possible. . . . The idea is to not have to do the gory details of endless trade studies. We'll just choose a [starlight-suppression] technology that has been progressing since the dawn of TPF as an idea, whichever one it is, and we'll throw our weight behind it, and if it fails, we walk away. You have to be willing to take risks. I know you've been talking to the guys at Space Telescope Science Institute, and they are my friends and they support this, but this is obviously

not the approach they're taking. You can't take risks like this in federally funded space science, and as a result, that model of building big,
complicated things is not very efficient. But if you're a private venture,
you do it your own way, on your own time and your own money, and
you bear the risk. That can make things smaller, more specialized, and
more affordable."

And what of the third path? I asked. Did Seager have a get-rich-
quick scheme I didn't know about?

She smiled. "This sounds like a joke, but it's very serious: mining
asteroids. If that happens in thirty, forty years, I'll be too old to run TPF,
but at least I'd have the money to make it personally happen." Seager
had signed on as a scientific advisor with a new venture, Planetary Resources, Inc., which would publicly debut two months after our conversation. The company was cofounded by two influential entrepreneurs of
the emerging private spaceflight industry, Eric Anderson and Peter
Diamandis; among its investors were Eric Schmidt and Larry Page of
Google, and the billionaire space tourist and software developer Charles
Simonyi. Other than Seager, its advisors included the Hollywood film-
maker and deep-ocean explorer James Cameron and a former U.S. Air
Force chief of staff, General T. Michael Moseley. The business plan
was, at its core, quite simple: locate and extract valuable resources from
near-Earth asteroids, many of which are thought to contain deposits of
platinum and other rare metals valued at trillions of dollars based on existing market prices. If, against long odds, the venture eventually proved
successful, its core team stood to net multibillion-dollar profits.

Planetary Resources planned to begin by building and launching
small space telescopes, both to remotely "prospect" asteroids and to sell
observing time to public and private parties. The next steps involved
creating a low-cost interplanetary communications network and sending fleets of nimble robotic spacecraft to rendezvous with the most
promising asteroids for closer inspection and eventual recovery of their
rich resources. Water and other volatiles could be processed into rocket
fuel, allowing the creation of orbital fuel depots, space-based gas

stations to supply paying customers. Platinum-group metals would be imported to Earth, where they could be used to vastly expand the consumer market for computing devices and renewable energy. Additional revenue would come from third-party licensing of the interplanetary comm network and the technology for cheap interplanetary spacecraft. Seager would provide her expertise in building small telescopes and on-orbit communications, acquired via her work with nanosatellites, as well as her access to MIT's community of researchers and her knowledge of remote photometric and spectroscopic observations. She saw the venture as part of a broader strategy for aiding the expansion of Earth's economic sphere into the rest of the solar system, and, someday, beyond.

"People forget that right now [space science] is considered a luxury field," she said. "It's not seen as an obligation like fighting poverty, or trying to cure AIDS or cancer, or combating global warming. We really can't just expect the government to do it for us; we very well may have to do it on our own, and having a robust commercial space industry can only help."

Later, she introduced me to the more concrete but no less futuristic portion of her plans: the young members of the various high-level MIT research groups she mentored, directed, and supervised. Some— Diana Valencia, Renyu Hu, Brice Demory, Vlada Stamenkovic—were already considered rising stars in their subfields, and had come from across Europe, Asia, and South America to work with Seager. Others— Becky Jensen-Clem, Christopher Pong, Mary Knapp, Matt Smith— were home-grown graduate or undergraduate students at MIT, earlier in their trajectories toward eminence and distinction. Each filled a crucial role in Seager's efforts toward studying exoplanets or building spacecraft. "They're almost like my extended family," Seager told me after I had met her young protégés. "They're another part of the legacy. They will grow up and fly away and populate the world with the next generation of great work in exoplanet atmospheres and interiors. . . . If I don't find the biosignatures, maybe some of them will."

In the evening, we took a train from Cambridge back to Seager's

Concord home. The house was a spacious three-story affair with a cozy screened-in porch and large backyard ringed with trees. Inside, Max and Alex greeted us from the living-room floor, sprawled on their bellies, brown-haired and barefoot, assembling Legos and scribbling in coloring books. Their babysitter packed her things and bid us goodnight. Seager pulled a sheet of poster board and some index cards from a nearby pile of papers. It was a game she had made with her children; "ALIENOPOLY" was handwritten in block letters across the poster board, above an image of a smiling, slug-like alien with eyes on stalks. Instead of buying Boardwalk and Park Place, players could individually purchase the planets of the Gliese 581 system, or the worlds of Alpha Centauri. Rolling the dice might place you on a wormhole, allowing you access to anywhere on the board, or could subject you to the indignity of an alien abduction and quarantine aboard a UFO. Seager excused herself to assemble a dinner of baked chicken, rice pilaf, and artichoke hearts in the cornflower-blue tiled kitchen, leaving me with the boys, who showed little interest in discussing their mother's work or playing Alienopoly.

"Do you like *Star Wars*?" they chirped in near-unison. I nodded. They exchanged a meaningful glance. Alex ran to a nearby couch, produced three toy light sabers from between the cushions, and padded back to thrust one into my hand. "Defend yourself, Darth Vader!" Max cried, raising his weapon. In two forty-five-minute bloodbaths before and after dinner, despite my Sith training I died many deaths at the hands of the young Jedis, repeatedly disemboweled, dismembered, and decapitated in a flailing whirlwind of gangly arms and Day-Glo plastic. Finally, well past their bedtime, Max and Alex reluctantly trudged upstairs to their beds. Seager and I stayed downstairs to chat over glasses of red wine while she did laundry.

She seemed puzzled by her sons' preference for swashbuckling space opera rather than the breathtaking realities she and her colleagues were unearthing every day. "Do you have any idea why they are so fixated on *Star Wars*?" she asked after a time.

I didn't, really, and mumbled about cultural archetypes in folklore and the "hero's journey" of Joseph Campbell, the timeless fantasy of frontiers and creatures that, while exotic, still bore some comforting semblance of familiarity.

"Maybe," she said, looking quizzical. "I don't really know what it is, why people do what they do." For a moment I lost the thread of our conversation. I was thinking of stories Seager had once told me, of Max pretending he was a visitor to Earth from an alien planet, and of Alex proclaiming he would become an astronaut to explore the Earth-like worlds his mother would discover. I remembered my own childhood, imagining I might someday visit a comet or be suddenly swept into the sky by a UFO to travel to another galaxy. Every child exists in a realm of infinite possibility, dreaming of other worlds, other lives—and of being singular and special enough to somehow reach and inhabit them. Whether it was unfulfilled potentials or harsh realities that made so many of those dreams fade at childhood's end, I couldn't say.

"I know I'm not going to live to see people travel to an exoplanet," Seager was saying. "But I can still make the maps. What happens after that is beyond what I can really comment on. Could a civilization, if they wanted to, marshal their resources to go to the nearest stars? I think that is within our reach." She excused herself to change a load of laundry, returning after a moment to talk about her new work in biosignatures.

Along with her collaborator William Bains, a UK-based biochemist, Seager was cooking up plausible alien worlds far different from Earth, trying to quantify what sorts of biospheres they might have and which sorts of biosignature gases could build up in their atmospheres. She and Bains wanted to assemble a catalog of potential varieties and scenarios, starting from the viewpoint that while Earth-like planets might be rare, planetary life was not.

She talked of anoxic "slime worlds," with oceans covered in great mats and blooms of biomass pumping out methane or hydrogen sulfide, and of gas-blanketed greenhouse worlds far from their stars, where creatures gained energy not from splitting water but from combining hydrogen and nitrogen to make ammonia. On warm ocean planets sheathed in crushingly thick atmospheres, she envisioned life existing deep down in bubbling aerosols formed by the fluid, turbulent interface of sea and sky, twilight-blue worlds where organisms could effortlessly swim and fly between air and water intermixed at exactly equal densities. Figuring out how such things could manifest in a future TPF-style telescope required simulating not one planetary surface and atmosphere, but millions, each with a different set of thermodynamically plausible assumptions that would affect the generation and visibility of biosignatures. As usual for Seager's work, some critics in the field seemed to think her new research was too futuristic to be useful—why go to such great lengths to clarify biosignatures for planetary environments that might not exist and that could only be observed using telescopes that might never be built?

"All this work is purely to make us ready for an eventual interpretation of unclear observations," she parried, sipping her wine. "And some of this will start happening soon. First we'll be able to look at some transiting super-Earths around nearby, quiet M-dwarfs, the ones with puffy, extended atmospheres that [the James Webb Space Telescope] and maybe even telescopes on the ground will be able to probe. But after that we won't have infinite chances to find our Earth twin. Anything launched in our lifetimes will probably only be able to look at the nearest hundred stars or so, and that's all we'll have. So if we don't have any Earth twins to survey in that space, how will we ever recognize any biosignature gases? Well, maybe we won't—unless we do this sort of extensive first-principles analysis. We could be very humbled by our first glimpses of super-Earth atmospheres."

I noted that she had forged her career out of being ahead of the curve, looking farther over the horizon than her peers. What did she think the future held? With the dream of a TPF deferred, what would sustain the exoplanet boom in the interim? Maybe after two decades of wild expansion, I offered, the field was on the verge of a dot-com–style crash.

Seager took a longer sip of wine, and rolled the glass stem between her fingers, thinking. "This really crystallized for me at the Next 40 Years [conference]," she began. "This field was built by rogue pioneers. It's a selection effect—you wouldn't have gotten into this if you weren't tough and belligerent and willing to take big risks and have high standards. Now the quality control is decreasing because so many people are flowing in. There are a ton of people who now appear to all be doing basically the same thing, with minor incremental variations. Everyone agrees we should map the nearby stars, but we don't have unity on the best approach." She bemoaned what she saw as a growing excess of sloppy theoretical work and observational results that added little to the field. "How many more transiting hot Jupiters do we really need?" she wondered. "Maybe we need more, but maybe we don't. I'm not the best to say."

But the lack of unity had its advantages, she went on. "Exoplanets are a bit like a dot-com bubble, yeah. But this will be like a hundred-year bubble, something that lasts a long time. It's always been true in astronomy that anytime you have a new technology and a new telescope, a new field opens up. It happened with high-precision spectrographs and radial velocity. It's happening with Kepler and transits. I see it like waves. You had an RV wave, it's now playing a supporting role, but it may have a resurgence later if most stars have Earths. Right now you have Kepler transits driving a wave. You'll have other waves later. The direct-imaging wave is coming. TPF is coming. They all overlap, but they all started at different times, so they have different phases. . . . The bubble will last because you can't have everything at the same time. Maybe if we had billions of dollars and did them all at once it would burn out. Because of the graduated opening of things, it's going to go on. Even once we find and directly image potential Earths, people will want to keep getting better resolution of their atmospheres and surfaces. But it won't last forever, it's true."

She finished her wine and checked the time. It was past 11:00 p.m., and I needed to catch the train back to Cambridge. Before I left, though, she said she wanted to show me something. We walked upstairs, past Max and Alex sleeping in their beds, and into a small study with bookshelves and a couch. She opened a nearby closet and retrieved several framed, yellowing photographs, laying them on the couch for display. Most were of Wevrick: pulling a canoe through rushing water, bestride a boulder on a massive esker's summit, haggard and dignified after a long day's hike. There he was, gazing into Seager's eyes as they embraced on their wedding day. He wore a black suit jacket and tie, and she was all in white, with pale flowers in her hair. Others showed plumes of undulating black smoke smearing out the blue sky above a shoreline of charred, dead trees, and sunlight shining off mist above pounding rapids.

"I used to be really into photography," Seager softly remarked. "On our trips, Mike would cook the food, and I would take the pictures. . . . I don't try to focus on my husband's death. But, you know, he died, and it was a major blow. It's having a huge ripple effect. Now I'm trying to live more purposefully. My children, my mentees, my students. I'm trying harder than ever before to motivate people to cut through the clutter and reach their dreams—myself included."

I met Seager in her office the next morning and accompanied her into the late afternoon through a blur of meetings, phone calls, and classes. Every hour a different part of her brain pivoted to a new set of problems, smoothly slewing from advising postdocs about their exoplanet research projects, to discussing the finer points of satellite thermal control with her graduate group, to giving project-management advice to undergraduate engineering students. By the evening I was exhausted, but Seager seemed inexhaustible—we separated for a couple of hours for her to take the qualifying exams for a ham radio license, which she wanted in relation to her interest in communications satellites. Dinner followed, sushi at a campus restaurant. Before parting ways, we backtracked through the night to the Green Building and took the elevator up to her seventeenth-floor office, where she had forgotten a bag. In the window of the unlit room, the city lights glinted off the dark rippling river, and for a moment we seemed to be floating in deep space, looking down on a galaxy's countless glittering stars. She looked up from rummaging around her desk and paused, bag in hand, entranced by the view.

"I love the skyline," she said, her back to me. "It gets me every time. The river. The sky. The light. It's a big part of my life, actually, the view and how it changes. How night falls. I look out and think about all the people, how the world fits together, the continuum of light. Day fades into night and then night fades back into day. Nature sets the path, but we do have some control. We're the product of millions of years of evolution, but we don't have any time to waste. That's what I've learned from death." Her voice broke and quavered, then through tears regained its strength. "Death made me realize how worthless most things are, yeah? Nothing else is meaningful; it supersedes everything. I have lost tolerance for things without meaning. There is no time for them. Does that make sense?"

In the darkness, a memory of a framed photograph she had shown me the night before swam unbidden to my mind. It was a rare shot snapped by Wevrick, taken from a height. A vast expanse of yellow grass and stunted trees swept out to the shores of a nameless lake, which stretched all the way to a treeless esker horizon. In the foreground, a lone point-like figure bent under an arc of red, casting a long shadow in the golden light of the Sun. It was Seager, steadfastly hauling the heavy canoe through the inhospitable terrain of the transition zone. The low-resolution photo did not convey whether the difficult portage was nearing its end or only just beginning. In the distance, the Barren Lands rolled on.

ACKNOWLEDGMENTS

This book was a long time coming, and a long list of people helped it along the way. I am thankful to Courtney Young for her belief in me, and to Emily Angell and Annie Gottlieb for sharpening my prose with their editorial oversight. My agent, Peter Tallack, has provided critical support from start to finish, and deserves much praise.

I am indebted to several members of my family for their moral and financial support. Were it not for the generosity of my parents, Mike and Pam Billings, and my grandparents, Bruce and Jo Hannaford, this book would not exist. My sister Carolyn and her husband Matt Tapie provided indulgent conversations and a roof over my head on some research trips. Most of all, I wish to thank my wife, Melissa Lherisson Billings, for her unwavering support, patience, and love.

My thanks to Adam Bly and the Seed Media Group editorial team for enabling some of my earliest encounters with a few key sources. Similarly, by inviting me to attend several of its annual symposia, the Miller Institute for Basic Research in Science at the University of

California, Berkeley, proved invaluable for incubating portions of this book. I am particularly grateful to Kathryn Day, Raymond Jeanloz, and Michael Manga for their kindness to me during my visits. Maggie Koerth-Baker, the science editor of Boingboing.net, was also instrumental in helping this project get off the ground in its formative stages. Nadia Drake deserves a very special thanks for arranging my meeting with her father.

For their friendship, advice, and encouragement, I must recognize: Evan Lerner, T. J. Kelleher, Paul Gilster, Joshua Roebke, Eric Weinstein, Jon Bardin, Ken Chang, Andrew Fullerton, Christopher Xu, Josh Chambers, George Musser, Carl Zimmer, and E.J. of Nevada County.

Over the years, a great many people generously answered my questions in ways that directly or indirectly informed this book. That said, any errors in this volume are entirely my own. I owe the following sources many thanks for their time and expertise:

Roger Angel, Guillem Anglada-Escudé, Mike Arthur, William Bains, Natalie Batalha, Charles Beichman, David Bennett, Michael Bolte, Xavier Bonfils, Alan Boss, John Casani, Webster Cash, John Chambers, Phil Chang, David Charbonneau, Nick Cowan, Paul Davies, Drake Deming, Frank Drake, Alan Dressler, Michael Endl, Debra Fischer, Kathryn Flanagan, Eric Ford, Colin Goldblatt, Mark Goughan, Jeff Greason, John Grunsfeld, Javiera Guedes, Olivier Guyon, Robin Hanson, Tori Hoehler, Andrew Howard, Jeremy Kasdin, Jim and Sharon Kasting, Heather Knutson, Antoine Labeyrie, David Latham, Greg Laughlin, Doug Lin, Jonathan Lunine, Kevin McCartney, Claudio Maccone, Bruce Macintosh, Geoff Marcy, John Mather, Greg Matloff, Michel Mayor, Vikki Meadows, Jon Morse, Matt Mountain, Phil Nutzman, Ben Oppenheimer, Bob Owen, Ron Polidan, Marc Postman, Sean Raymond, Dimitar Sasselov, Jean Schneider, Sara Seager, Michael Shao, Seth Shostak, Rudy Slingerland, Chris Smith, Rémi Soummer, David Spergel, Alan Stern, Peter Stockman, Jill Tarter, Philippe Thébault, Wes Traub, Michael Turner, Stéphane Udry, Steve Vogt, Jim Walker, Bernie Walp, Andrew Youdin, and Kevin Zahnle.

SELECTED FURTHER READING AND NOTES

CHAPTER 1:
Looking for Longevity

Ronald N. Bracewell, *The Galactic Club: Intelligent Life in Outer Space* (San Francisco: W. H. Freeman, 1974).

Giuseppe Cocconi and Philip Morrison, "Searching for Interstellar Communications," *Nature*, vol. 184 (1959), pp. 844–46.

Frank Drake and Dava Sobel, *Is Anyone Out There? The Scientific Search for Extraterrestrial Intelligence* (New York: Delacorte Press, 1992). I quote Drake from page 27.

Stanislaw Lem, *Summa Technologiae* (Minneapolis: University of Minnesota Press, 2013; first edition, 1964). Translated by Joanna Zylinska, this is the first complete English translation of Lem's prescient classic on cosmic evolution.

J. P. T. Pearman, "Extraterrestrial Intelligent Life and Interstellar Communication: An Informal Discussion," in *Interstellar Communication*, A. G. W. Cameron, ed. (New York: W. A. Benjamin, 1963), pp. 287–93.

Iosif Shklovskii and Carl Sagan, *Intelligent Life in the Universe* (San Francisco: Holden-Day, 1966).

Walter Sullivan, *We Are Not Alone: The Continuing Search for Extraterrestrial Intelligence*, rev. ed. (New York: Dutton, 1993).

Otto Struve, "Proposal for a Project of High-Precision Stellar Radial Velocity Work," *The Observatory*, vol. 72 (1952), pp. 199–200.

CHAPTER 2:
Drake's Orchids

J. D. Bernal, *The World, the Flesh, and the Devil: An Enquiry into the Future of the Three Enemies of the Rational Soul* (London: Kegan Paul, Trench, Trübner, 1929).

Paul Davies, *The Eerie Silence: Renewing Our Search for Alien Intelligence* (New York: Houghton Mifflin Harcourt, 2010).

Frank Drake, "Stars as Gravitational Lenses," in *Bioastronomy—The Next Steps*, G. Marx, ed., Astrophysics and Space Science Library, vol. 144 (Dordrecht: Kluwer Academic Publishers, 1988), pp. 391–94.

Frank Drake and Dava Sobel, *Is Anyone Out There? The Scientific Search for Extraterrestrial Intelligence* (New York: Delacorte Press, 1992). Drake's calculation of how many boxes of corn flakes the Arecibo Observatory radio dish could hold appears on pages 73–74.

Von R. Eshleman, "Gravitational Lens of the Sun: Its Potential for Observations and Communications Over Interstellar Distances," *Science*, vol. 205 (1979), pp. 1133–35.

Paul Gilster, *Centauri Dreams: Imagining and Planning Interstellar Exploration* (New York: Springer, 2004).

Hans Moravec, *Mind Children: The Future of Robot and Human Intelligence* (Cambridge, MA: Harvard University Press, 1988).

Peter D. Ward and Donald Brownlee, *Rare Earth: Why Complex Life Is Uncommon in the Universe* (New York: Springer, 2000).

CHAPTER 3:
A Fractured Empire

Guillem Anglada-Escudé et al., "A Planetary System around the Nearby M Dwarf GJ 667C with At Least One Super-Earth in its Habitable Zone," *The Astrophysical Journal Letters*, vol. 751 (2012), pp. L16–.

Lee Billings, "G is for Goldilocks," Seedmagazine.com, October 1, 2010. http://seedmagazine.com/content/article/g_is_for_goldilocks/.

Xavier Bonfils et al., "The HARPS Search for Southern Extra-solar Planets XXXI. The M-dwarf Sample," *Astronomy & Astrophysics*, vol. 549 (2013), pp. 109–.

Alan Boss, *The Crowded Universe: The Search for Living Planets* (New York: Basic Books, 2009).

Xavier Delfosse et al., "The HARPS Search for Southern Extra-solar Planets XXXV. Super-Earths around the M-dwarf Neighbors Gl433 and Gl667C," arXiv preprint (2012).

Bruce Dorminey, *Distant Wanderers: The Search for Planets Beyond the Solar System* (New York: Springer, 2001).

Thierry Forveille et al., "The HARPS Search for Southern Extra-solar Planets XXXII. Only 4 planets in the Gl~581 system," arXiv preprint (2011).

Philip C. Gregory, "Bayesian Re-analysis of the Gliese 581 Exoplanet System," arXiv e-print (2011).

Ray Jayawardhana, *Strange New Worlds: The Search for Alien Planets and Life Beyond Our Solar System* (Princeton: Princeton University Press, 2011).

Marc Kaufman, *First Contact: Scientific Breakthroughs in the Hunt for Life Beyond Earth* (New York: Simon & Schuster, 2011).

Michael D. Lemonick, *Other Worlds: The Search for Life in the Universe* (New York: Simon & Schuster, 1998).

Tim Stephens, "Newly discovered planet may be first truly habitable exoplanet," University of California, Santa Cruz, Newscenter, September 29, 2010. http://news.ucsc.edu/2010/09/planet.html.

Otto Struve, "Astronomers in Turmoil," *Physics Today*, vol. 13 (1960), p. 18.

Mikko Tuomi, "Bayesian Re-analysis of the Radial Velocities of Gliese 581. Evidence in Favour of Only Four Planetary Companions," arXiv preprint (2011).

Steven S. Vogt et al., "The Lick-Carnegie Exoplanet Survey: A 3.1 M_Earth Planet in the Habitable Zone of the Nearby M3V Star Gliese 581," *The Astrophysical Journal*, vol. 723 (2010), pp. 954–65.

Steven S. Vogt, R. Paul Butler, and Nader Haghighipour, "GJ 581 Update: Additional Evidence for a Super-Earth in the Habitable Zone," *Astronomische Nachrichten*, vol. 333 (2012), pp. 561–75.

CHAPTER 4:
The Worth of a World

Fred Adams and Greg Laughlin, *The Five Ages of the Universe: Inside the Physics of Eternity* (New York: Free Press, 1999).

John D. Barrow and Frank J. Tipler, *The Anthropic Cosmological Principle* (New York: Oxford University Press, 1986).

Marcia Bartusiak, *The Day We Found the Universe* (New York: Pantheon, 2009).

Lee Billings, "Cosmic Commodities: How much is a new planet worth?" Boingboing.net, February 3, 2011. http://boingboing.net/2011/02/03/cosmic-commodities-h.html.

Bill Bryson, *A Short History of Nearly Everything* (New York: Broadway Books, 2003).

Thane Burnett, "Wanna buy the Earth? It'll cost you $5 quadrillion," *Toronto Sun*, March 1, 2011. http://www.torontosun.com/news/columnists/thane_burnett/2011/03/01/17455846.html.

Robert Costanza et al., "The value of the world's ecosystem services and natural capital," *Nature*, vol. 387 (1997), pp. 253–60.

Michael J. Crowe, ed., *The Extraterrestrial Life Debate, Antiquity to 1915: A Source Book* (Notre Dame: University of Notre Dame Press, 2008). Daily Mail Reporter, "Earth is worth £3,000 trillion, according to scientist's new planet valuing formula," MailOnline.com, February 28, 2011. http://www.dailymail.co.uk/sciencetech/article-1361145/Earth-worth-3-000-trillion-according-scientists-new-planet-valuing-formula.html.

Steven J. Dick, *The Biological Universe: The Twentieth-Century Extraterrestrial Life Debate and the Limits of Science* (Cambridge, UK: Cambridge University Press, 1996).

Stephen Greenblatt, *The Swerve: How the World Became Modern* (New York: W. W. Norton, 2011).

Alan H. Guth, *The Inflationary Universe: The Quest for a New Theory of Cosmic Origins* (New York: Perseus Books, 1997).

Arthur Koestler, *The Sleepwalkers: A History of Man's Changing Vision of the Universe* (New York: Macmillan, 1959).

D. G. Korycansky, Gregory Laughlin, and Fred C. Adams, "Astronomical engineering: a strategy for modifying planetary orbits," *Astrophysics and Space Science*, vol. 275 (2001), pp. 349–66.

Greg Laughlin, "Too cheap to meter," *systemic*, March 12, 2009. http://oklo.org/2009/03/12/too-cheap-to-meter/.

Lucretius, *On the Nature of Things* (Newburyport, MA: Focus Publishing, 2003). I quote from pages 59 and 60 of Walter Englert's excellent translation.

Carl Sagan, *Cosmos* (New York: Random House, 1980).

Alex Vilenkin, *Many Worlds in One: The Search for Other Universes* (New York: Hill and Wang, 2006).

CHAPTER 5:
After the Gold Rush

Lee Billings, "The Long Shot," Seedmagazine.com, May 19, 2009. http://seedmagazine.com/content/article/the_long_shot/.

Ray Bradbury, *The Martian Chronicles* (Garden City, NY: Doubleday, 1950). Laughlin's quote comes from "The Naming of Names" section of Bradbury's classic, p. 136.

Xavier Dumusque et al., "An Earth-mass planet orbiting α Centauri B," *Nature*, vol. 491 (2012), pp. 207–11.

Freeman John Dyson, "Search for Artificial Stellar Sources of Infrared Radiation," *Science*, vol. 131 (1960), pp. 1667–68.

Paul Gilster, "New Search for Centauri Planets Begins," *Centauri Dreams*, December 2, 2009. http://www.centauri-dreams.org/?p=10489.

Javiera Guedes et al., "Formation and Detectability of Terrestrial Planets Around Alpha Centauri B," *The Astrophysical Journal*, vol. 679 (2008), pp. 1582–87.

Edward Singleton Holden, *A Brief Account of the Lick Observatory of the University of California (1895)* (Whitefish, MT: Kessinger Publishing, 2010).

Andrew W. Howard et al., "The Occurrence and Mass Distribution of Close-in Super-Earths, Neptunes, and Jupiters," *Science*, vol. 330 (2010), pp. 653–55.

Andrew W. Howard et al., "Planet Occurrence within 0.25 AU of Solar-type Stars from Kepler," *The Astrophysical Journal: Supplement Series*, vol. 201, no. 2 (2012).

John McPhee, *Assembling California* (New York: Farrar, Straus and Giroux, 1993).

Isaac William Martin, *The Permanent Tax Revolt: How the Property Tax Transformed American Politics* (Stanford: Stanford University Press, 2008).

Tom Murphy, "Galactic-Scale Energy," *Do the Math*, July 12, 2011. http://physics.ucsd.edu/do-the-math/2011/07/galactic-scale-energy/.

Kevin Starr, *California: A History* (New York: Modern Library, 2005).

Philippe Thébault, Francesco Marzari, and Hans Scholl, "Planet formation in the habitable zone of alpha Centauri B," *Monthly Notices of the Royal Astronomical Society Letters*, vol. 393 (2009), pp. L21–L25.

Various authors, *2063 A.D.* (San Diego: General Dynamics Astronautics, 1963).

Various authors, The Lick Observatory Historical Collections Project, http://collections.ucolick.org/archives_on_line/.

CHAPTER 6:
The Big Picture

Paul J. Crutzen, "Geology of mankind," *Nature*, vol. 415 (2002), p. 23.

Paul J. Crutzen and Eugene F. Stoermer, "The 'Anthropocene,'" *Global Change Newsletter*, vol. 41 (2000), pp. 17–18.

Christian de Duve, *Vital Dust: Life as a Cosmic Imperative* (New York: Basic Books, 1995).

James Hansen, *Storms of My Grandchildren: The Truth About the Coming Climate Catastrophe and Our Last Chance to Save Humanity* (New York: Bloomsbury, 2009).

Andrew H. Knoll, *Life on a Young Planet: The First Three Billion Years of Evolution on Earth* (Princeton: Princeton University Press, 2003).

James Lovelock, *The Vanishing Face of Gaia: A Final Warning* (New York: Basic Books, 2009).

Seamus McGraw, *The End of Country: Dispatches from the Frack Zone* (New York: Random House, 2011).

John McPhee, *Basin and Range* (New York: Farrar, Straus and Giroux, 1981). I quote from McPhee's embodiment of the geological timescale on page 127, which comes in the midst of a more general discussion of deep time that, in tone and outlook, has greatly influenced by own thoughts on the subject as presented in this book.

Oliver Morton, *Eating the Sun: How Plants Power the Planet* (New York: HarperCollins, 2008).

Andrew Revkin, *Global Warming: Understanding the Forecast* (New York: Abbeville Press, 1992). Revkin presaged Crutzen and Stoermer's "Anthropocene" by eight years; he called the dawning geological era the "Anthrocene."

William F. Ruddiman, *Plows, Plagues, and Petroleum: How Humans Took Control of Climate* (Princeton: Princeton University Press, 2005).

J. William Schopf and Cornelis Klein, eds., *The Proterozoic Biosphere: A Multidisciplinary Study* (Cambridge, UK: Cambridge University Press, 1992).

Various authors, Marcellus Center for Outreach and Research, http://www .marcellus.psu.edu. This website contains a wealth of information about gas-shale fracking in Pennsylvania and includes detailed publication lists and geological maps.

Jan Zalasiewicz et al., "Stratigraphy of the Anthropocene," *Philosophical Transactions of the Royal Society A: Mathematical, Physical and Engineering Sciences*, vol. 369 (2011), pp. 1036–55.

CHAPTER 7:
Out of Equilibrium

John A. Baross et al., *The Limits of Organic Life in Planetary Systems* (Washington, DC: National Academies Press, 2007). This report is available online at http://www.nap.edu/catalog.php?record_id=11919.

Stephen H. Dole and Isaac Asimov, *Planets for Man* (New York: Random House, 1964). A condensed and popularized version of Dole's RAND Corporation Research Study, *Habitable Planets for Man*.

L. Kaltenegger, S. Udry, and F. Pepe, "A Habitable Planet around HD 85512?" arXiv preprint (2011).

James Kasting, *How to Find a Habitable Planet* (Princeton: Princeton University Press, 2009). I recommend this excellent book to those readers wishing to seek out the scholarly papers mentioned in Chapter 7, as well as any readers wishing for a readable-yet-rigorous take on planetary habitability.

Ravi Kumar Kopparapu et al., "Habitable Zones Around Main-Sequence Stars: New Estimates," arXiv preprint (2013).

Charles H. Langmuir and Wally Broecker, *How to Build a Habitable Planet: The Story of Earth from the Big Bang to Humankind*, revised and expanded ed. (Princeton: Princeton University Press, 2012).

J. E. Lovelock, *Gaia: A New Look at Life on Earth* (New York: Oxford University Press, 1979).

Stephen H. Schneider and Penelope J. Boston, eds., *Scientists on Gaia* (Cambridge, MA: MIT Press, 1992).

Peter Ward, *The Medea Hypothesis: Is Life on Earth Ultimately Self-Destructive?* (Princeton: Princeton University Press, 2009).

Peter D. Ward and Donald Brownlee, *The Life and Death of Planet Earth: How the New Science of Astrobiology Charts the Ultimate Fate of Our World* (New York: Times Books, 2003).

CHAPTER 8:
Aberrations of the Light

Norman R. Augustine et al., *Seeking a Human Spaceflight Program Worthy of a Great Nation* (Washington, DC: NASA, 2009). This report is available online at http://www.nasa.gov/pdf/396093main_HSF_Cmte_FinalReport.pdf.

Lee Billings, "The telescope that ate astronomy," *Nature*, vol. 467 (2010), pp. 1028–30.

Alan Boss, *The Crowded Universe: The Search for Living Planets* (New York: Basic Books, 2009).

Kenneth G. Carpenter et al., "OpTIIX: An ISS-based Testbed Paving the Roadmap toward a Next Generation, Large Aperture UV/Optical Space Telescope," June 20, 2012. uvastro2012.colorado.edu/Presentations /KennethCarpenter.pdf. This material was presented at the "UV Astronomy: HST and Beyond Conference" hosted by the University of Colorado, Boulder, from June 18–21, 2012.

Andrew Chaikin, *A Man on the Moon: The Voyages of the Apollo Astronauts* (New York: Viking Penguin, 1994).

Daniel S. Goldin, "NASA in the Next Millennium," January 17, 1996. http:// home.fnal.gov/~annis/digirati/otherVoices/goldin.AAS.html. This is the transcript of Goldin's speech at the 187th American Astronomical Society meeting in San Antonio, Texas, and is the source for Goldin's quotes in this chapter.

Greg Klerkx, *Lost in Space: The Fall of NASA and the Dream of a New Space Age* (New York: Pantheon, 2004).

John M. Logsdon, *John F. Kennedy and the Race to the Moon* (New York: Palgrave Macmillan, 2010).

Matt Mountain, Untitled Presentation at TEDxMidAtlantic 2010, November 5, 2010. http://www.youtube.com/watch?v=_4qO4GjyyUI. This public talk contains many of the same points and visualizations that Mountain employed during our discussions.

Michael J. Neufeld, *Von Braun: Dreamer of Space, Engineer of War* (New York: Knopf, 2007).

N. A. Rynin, author and editor, *Interplanetary Flight and Communication, Vol. III, No. 7: K. E. Tsiolkovskii: Life, Writings, and Rockets* (Jerusalem: Israel Program for Scientific Translations, 1971). Rynin originally self-published this volume in Leningrad in 1931. I have drawn Tsiolkovsky's quotes from pages 3, 7, 30, and 31.

Robert Zimmerman, *The Universe in a Mirror: The Saga of the Hubble Space Telescope and the Visionaries Who Built It* (Princeton: Princeton University Press, 2008).

CHAPTER 9:
The Order of the Null

C. A. Beichman, N. J. Woolf, and C. A. Lindensmith, eds., *The Terrestrial Planet Finder (TPF): A NASA Origins Program to Search for Habitable Planets* (Pasadena: NASA-Caltech Jet Propulsion Laboratory, 1999). This report is available online at http://exep.jpl.nasa.gov/TPF/tpf_book/index.cfm.

Michael Belfiore, *Rocketeers: How a Visionary Band of Business Leaders, Engineers, and Pilots Is Boldly Privatizing Space* (Washington, DC: Smithsonian, 2007).

Lee Billings, "Let There Be Light," Seedmagazine.com, November 17, 2009. http://seedmagazine.com/content/article/let_there_be_light/.

Roger D. Blandford et al., *New Worlds, New Horizons in Astronomy and Astrophysics* (Washington, DC: National Academies Press, 2010). This report is available online at http://www.nap.edu/catalog.php?record_id=12951.

Chris Dubbs and Emeline Paat-Dahlstrom, *Realizing Tomorrow: The Path to Private Spaceflight* (Lincoln: University of Nebraska Press, 2011).

Charles Elachi et al., *A Road Map for the Exploration of Neighboring Planetary Systems* (Pasadena: NASA-Caltech Jet Propulsion Laboratory, 1996). This report is available online at http://exep.jpl.nasa.gov/exnps/toc.html.

James Kasting, Wesley Traub et al., *Terrestrial Planet Finder—Coronagraph (TPF-C) Flight Baseline Mission Concept* (Pasadena: NASA-Caltech Jet Propulsion Laboratory, 2009). This report is available online at http://exep.jpl.nasa.gov/TPF-C/TPFC-MissionAstro2010RFI-Final-2009-04-01.pdf.

Michael D. Lemonick, *Mirror Earth: The Search for Our Planet's Twin* (New York: Walker, 2012). In addition to providing a marvelous overview of the history and forefront of planet hunting, Lemonick's book also offers a detailed independent perspective on the events of Seager's "The Next 40 Years of Exoplanets" conference.

Jonathan Lunine et al., "Worlds Beyond: A Strategy for the Detection and Characterization of Exoplanets," Report of the ExoPlanet Task Force, Astronomy and Astrophysics Advisory Committee (Washington, DC: NSF /NASA, 2008). This report is available online at http://www.nsf.gov/mps/ast /aaac/exoplanet_task_force/reports/exoptf_final_report.pdf.

Christopher F. McKee, Joseph H. Taylor, Jr., et al., *Astronomy and Astrophysics in the New Millennium* (Washington, DC: National Academies Press, 2001). This report is available online at http://www.nap.edu/catalog.php ?record_id=9839.

Sara Seager et al., "The Next 40 Years of Exoplanets," MIT, May 27, 2011. http://seagerexoplanets.mit.edu/next40years.htm. Program speakers, photographs, and a web archive of the videotaped proceedings are available at this hyperlink.

W. A. Traub and B. R. Oppenheimer, "Direct Imaging of Exoplanets," in *Exoplanets*, Sara Seager, ed. (Tucson: University of Arizona Press, 2010).

Stephen C. Unwin et al., "Taking the Measure of the Universe: Precision Astrometry with SIM PlanetQuest," *Publications of the Astronomical Society of the Pacific*, vol. 120 (2008), pp. 38–88.

CHAPTER 10:
Into the Barren Lands

Freeman Dyson, *Disturbing the Universe* (New York: Harper & Row, 1979).

John S. Lewis, *Mining the Sky: Untold Riches from the Asteroids, Comets, and Planets* (New York: Perseus Books, 1996).

Gerard K. O'Neill, *The High Frontier: Human Colonies in Space* (New York: William Morrow, 1976).

Carl Sagan, *Pale Blue Dot: A Vision of the Human Future in Space* (New York: Random House, 1994). I quote Sagan from pages 6 and 7.

Carl Sagan et al., *Murmurs of Earth: The Voyager Interstellar Record* (New York: Random House, 1978).

Sara Seager, "Sixty Days in the Land of Little Sticks: Part 2: Nowleye and Kamilukuak Rivers, Casimir and Kasba Lakes," *Nastawgan, the Quarterly Journal of the Wilderness Canoe Association*, vol. 23, no. 3 (1996), pp. 1–7.

Sara Seager and Mike Wevrick, "Sixty Days in the Land of Little Sticks: Part 1: Cochrane, Thlewiaza, Little Partridge, and Kazan Rivers," *Nastawgan, the Quarterly Journal of the Wilderness Canoe Association*, vol. 23, no. 1 (1996), pp. 1–7.

Henry David Thoreau, *Walden; or, Life in the Woods* (Boston: Ticknor and Fields, 1854). I quote from the book's fifth section, entitled "Solitude."

Robert Zubrin, *Entering Space: Creating a Spacefaring Civilization* (New York: Tarcher, 1999).

INDEX